My Magic Circle

Poems by Julia Wen

The Commercial Press

Contents

Dedication iv

Foreword
Carson Wen v

Prologue
Ian Wen viii
Anna Qi xi

Preface
Chiu Ha Ying xiv
Sharon Evans xvii
Barbara Meynert xix
Edith Shih xxi
Amarantha Yip xxiii
Chris Wat xxv

Acknowledgements xxvii

A Circle of Endless Love
My Muse, My Light and My Love 2
For Anna 8
For My Son 12
For My Most Loving Husband 16
Birthdays for My Two Men 17
For My Dearest Son 21
Happy 21st Birthday, Darling 24

A Coronet of Priceless Gems
For Winnie 28
For Elizabeth 33
For Eunice 38

A Birthday Tribute to Ada   43
For Kathy   48
For Charmaine   53
For Susie   57
For Ruth   63
For Sharon   69
For Barbara   74
For Chris   80
A Birthday Tribute to Ha Ying   82
A Birthday Tribute to Maria   87
A Birthday Tribute to Amarantha   90
For Stephen at 70   95
For Edith   100
A Birthday Tribute to Jerry   102
A Birthday Tribute to Jimmy   106
A Birthday Tribute to Fai   112
A Birthday Tribute to Alan   118
A Birthday Tribute to Edith   121
A Birthday Tribute to Cindy   126
A Birthday Tribute to Elizabete   129
For Eliza   134
For My Philippine Angel   137
For Edith   141
For Ha Ying   143
For Stephen at 60   147
A Birthday Tribute to Virginia   151
To My Doctor Friend   156
For CN   160
For Shirley   165

A Garland of Cherished Friendship

For the Suns   170
Ode to My Women Friends   174
Happy Pearl Anniversary to Christine and HF   180

Ode to St Mary's   185
A Tribute to the Orange Party   188
A Thank You Tribute to My Comrades   192
For All My Mediator Friends with Love   195
Farewell My Friends, Fare Thee Well   197
Ode to Knowles Building   201
A Tribute to My Gang of Four   216

A Halo of Gentle Souls

A Tribute to Our Dearest Mary   220
To Suk Suk   223
In Loving Memory of Our Dearest Friend   227
Celebrating a Life Well Lived   230
Light a Candle   234

Dedication

I wish to dedicate this book to my husband, my confidant and soulmate, who is 90% inspiration, 90% patience, 90% encouragement, 90% indulgence, 90% motivating force and 90% guiding light.

They do not add up.

He multi-tasks.

Foreword

I am writing this Prologue for the fourth poetry collection by my wife Julia, who is the Love of my Life.

I was twice introduced to Julia in consecutive weeks in the '70s, first by Haruki (Alex) Shimbo, the then Hong Kong based representative of a Japanese client of mine who became a personal friend, and a week later by Tan Eng Chuan, a former schoolmate at Columbia, who became a big-time banker. Alex made the introduction as Julia was planning a Tokyo trip and Alex thought I, who was to relocate to Tokyo soon to represent my Hong Kong law firm, could be her local host there. Tan, whose wife is a close friend of Julia's sister Susanna, was doing a surprise matchmaking by inviting us both home for dinner.

After years of courtship which entailed the writing of a love poem on every Sunday and the delivery of same to Julia together with roses on every Monday, I managed to persuade Julia to tie the knot and we married in January 1983. Having gotten her hand, a wife to support and a family to raise, I stopped writing my weekly love poems after our wedding and devoted time to work hard, unknowingly developing myself into a workaholic. We live love by taking tender loving care of each other, and that has continued to this day. As for poetry, Julia lives it, breathes it and embodies it. For me to write her weekly love poems would be belittling the Poet in her. She Is Poetry.

Julia lives poetry by her purity. Despite her experiences in life with her past career in publication, media, government, academia and art, and friendships with senior decision makers, Julia's soul remains as pure as the youngest lily in a pond. Many friends of mine remarked that Julia 不吃人間煙火, and age has not mellowed her. She sees the world through the eyes of her inner child, the eyes that sparkle in awe and amazement as they see love, magic and beauty even in the most ordinary things. Experience and knowledge have not yielded cunning, nor has abundance corrupted her. She remains often astonished by how badly people take advantage of each other, when those are commonplace. Julia is fresh air and one cannot paint on or taint it.

Julia lives poetry by her compassion. She is touched by every kind gesture that was extended to her by friends and strangers, and repays those with love, sometimes with poems. She is pained by the suffering that she sees around the world, and investigates ways to make lives better for the under-privileged, the exploited and the destitute. She has committed to many service and philanthropist projects throughout the years, and has formulated, with the help of a team of social sciences specialists a "A Low-income Working Women Wellness Index in Hong Kong" in 2017, under the auspices of the Zonta Club of Hong Kong, to help these women benchmark themselves on emotional, psychological and social patterns and measurements and find options for self betterment.

Julia lives poetry by her intellectual curiosity. History awes her, philosophy enlightens her, literature enriches her, politics enthrals her and economics perplexes her. Some of our quality time together were spent on discussing topics that I may have some knowledge or experience to share, and she is such a good listener and keen inquisitionist. She is much influenced by my anti-imperialist stances, and has researched into the atrocities and exploitation that the British Empire has brought to their colonies in Africa, Asia and elsewhere. She also likes to listen to my account of the government's successful defence of the Hong Kong Dollar in 1998 against the plot of Soros and his collaborators to undermine it.

Julia lives poetry by her good taste. She has an eye for beauty which she applies to various aspects of her life – fashion, paintings and other works of art, lifestyle, cultural or simple everyday household wares. She's able to discern the exquisite essence of things without being affected by the unnecessary. While not a big spender by any sense, she does collect things of beauty, thus enriching my life as well. But above everything else she has good taste for friends, a taste that cannot be taught, bought or borrowed but one that comes from inherent appreciation and understanding. She has stayed amazingly young looking despite our growing years, and is a living example of the adage that age is a state of mind and style a state of heart.

Julia lives poetry by her wit. Hers is always one of joy, and such fun and laughter that I am reminded of the famous lines by William Blake: "when the green woods laugh with the voice of joy, and the dimpling stream runs laughing by; when the air does laugh with our merry wit, and the green hill laughs with the noise of it". She's an avid reader of English Literature and I think she might have somehow picked up humour from some of the works of those great masters. Her Oxford trained husband sadly does not live up to such standards and stays very boring by her measures. She always adds fun and laughter at social gatherings, even with lawyers and politicians, probably by taking them out of their shell by asking otherwise politically incorrect questions, sometimes terrifying me.

Julia lives poetry by her observant heart. No one is safe from being encapsulated, and succinctly described in her poems. She has the poetic sensitivity to detect and memorialize individual yearnings or pains, joys or frustrations, pride and prejudice, jealousies and generosities, changing moods, evolving social mores, complex relationships, radical social or intellectual movements, and the dawning of a post Western centric world. She listens, googles, and studies whatever new ideas or information that she has hitherto not known, and contextualises those in her writings. By doing so her poems do not have expiry dates.

Julia lives poetry by her loyalty. Despite her colonial education she is very patriotic. She even went to talk to protestors during Occupy Central and the 2019 social unrests by trying to tell them right from wrong, and surprisingly did not get beaten up. She is very loyal to her friends and family. The passing of some dear friends devastated her, and she wrote in memoriam poems for them even in the early hours of the darkness of the night that followed each passing, when grief overwhelmed her. She has written birthday, celebratory, and at times sentimental poems for her friends. She has so many I have lost count of them.

Julia lives poetry by her discipline. Though vivacious and fun loving, she is diligent and works herself very hard to deliver on goals that she sets for herself, whether in intellectual pursuits or community work. I have enjoyed seeing her at work, from curating art exhibitions, preparing and delivering lectures on modern English poetry or family mediation, organizing fundraising projects to writing art and poetry books. She takes extremely good care of myself, Ian and Anna, and puts our interests ahead of her own. In the proverbial last lifeboat for three scenario, she will be the one who insists on staying on the sinking ship, not that we are going to let her do it.

Thank you all for reading *The Magic Circle.* Like me, I hope you will discover and savour the poetry that is Julia.

Carson

Prologue

My mother once showed me a Haiku poem written to her by a friend:

Poet wife mother
Unique and Captivating
Julia, gift of love

The poem adeptly captures my mother's essence in 17 syllables in true Haiku style, but she is all that and much more.

My mother is a beautiful person. She's a loving and caring wife; an understanding and attentive mother; a kind, warm, considerate, thoughtful and high-spirited friend, a resourceful and committed volunteer worker and a true patriot.

My dad and I have always been the centre of her whole universe. She is never about herself – she's all about us, and she dotes on us; and now that my dad is not getting any younger, my mother has told me since at least fifteen years ago that dad is priority.

My mother loves life in its full splendour. Her favourite ways of 'enjoying life' are reading a good book over a glass of warm water, doing social community work and debating or having discussions with friends on art, culture, literature, lifestyle and other more serious topics like world history, economics, financial and trade summits and conflicts, and in recent years, political division and social unrest.

Nothing could make my mother happier than spending time with friends she enjoys and what numerous friends she has, and in such varieties too: from lawyers to doctors to journalists to musicians to mediators to therapists to academics to artists to civil servants to politicians; from Christians to Catholics to Buddhists to atheists; from friends known for over forty years to new ones she has just met during Hong Kong's darkest hours in 2019 when they worked together to render whatever little they could to help this city they so love. I thought I know most of her groups of friends and could recognize whoever she's referring to when I read her Section under "A Garland of Cherished Friendship" only to find out when I was helping her to install an app on her iPhone that there are so many other groups with such hilarious group names that I actually asked her to explain to me what they meant, and she most happily did.

These groups like "My Movie Trio", "The Four Little Flowers", "The Golden

Foursome", "The Three Musketeers", "My Super Mediator Comrades", "Video Production Troupe", "The Iron Foursome", "Fellow Travellers" or "Problem Solvers" have no poems included in this collection which gives me hope that there will be another volume before too long much as my mother keeps telling me that this will be her last project.

I enjoy reading her Section under "A Coronet of Priceless Gems" which contains the bulk of the collection. As I read every one of them, I realize why my mother values each and every one of them despite how unique and different one is from another, for each shares the same traditional Chinese core values of family harmony and solidarity, parental love, premium on education, filial piety and love and respect for authority.

I particularly like her "Ode to Knowles Building", the building where she spent her university days at the University of Hong Kong:

"When we were but 18,
the Chosen Ones,
each with an entry ticket
to the elite world of tomorrow..."

The poem gives me a rare glimpse of the era she was born into, the university life so different from my own, the city of Hong Kong under colonial rule, and the outside world, an epic poem spanning over a total of thirty-five years that future historians will find most informative and interesting.

My absolute favourite poem is the one she wrote for my father which is inspired by Sylvia Plath, the American poet. It's poignant and yet simple, and the last lines

"to weave a garland of love for you...
for you, for you, for you"

echoes most seamlessly with the first

"I am, I am I am"

My mother has indeed weaved a garland of love not only for my dad but for all of us. Her love for us is nothing short of an epic poem.

My mother has chosen four poems she wrote for me at different times of my growing up and I like her latest best. When I came to the part where my mother said that she now realizes that I am already grown up and could no longer be my light and asked if she could continue to be my soulmate and companion, I was very touched and my heart was filled with deep emotions of love and gratitude. I am particularly grateful for the beautiful poem she has written and included here for my wife Anna which moved both of us to tears.

"And now that you have joined us,
our family is blissfully complete."

The poem demonstrates the unselfish and generous love of a mother who has so much capacity to love. It's in her nature to love and care for others like she's born with knowing exactly how to do that well.

We are so blessed to have a loving mother like you.

We love you, mum.

Ian

Prologue

在律師圈，Magic Circle 作為頂尖律師事務所的代名詞，幾乎是每個年輕律師夢想進入的圈子，當馮月珊女士用 My Magic Circle 作為她的第四本詩集的標題時，最務實的法律和最浪漫的詩歌發生了碰撞，產生了奇妙的火花，這正如我們家的生活一樣，在一個有三位律師的家庭裏，馮女士－我婆婆、也是我媽媽－正是這律師圈裏的那一位浪漫詩人，她猶如一抹絢麗的色彩，豐富了生活的畫板。

詩的魅力，即使穿越千年，依然熠熠生輝，從"關關雎鳩，在河之洲"，到"明月幾時有，把酒問青天"，再到"眾裏尋他千百度，那人卻在燈火闌珊處"，讀詩，不僅是品味語言的魅力、欣賞音韻的和諧，更是探索詩人的心境、體悟情感的共鳴。錢穆先生曾說："我們學做文章，讀一家作品，也該從他筆墨去了解他的胸襟。我們不必要想自己成個文學家，只要能在文學裏接觸到一個較高的人生，接觸到一個合乎我自己的更高的人生。"此次讀 *My Magic Circle* 詩集，我想我們也不必要求自己成為一個詩人，只要能通過詩集接觸到不同的人生、認識有趣的靈魂、體會同樣的情感，那麼，這本詩集，已經可以為我們的心靈提供一個休憩之所。

My Magic Circle 詩集一共收集了 54 首詩，每一首詩要麼刻畫了一個人物、要麼描述了一段經歷，詩中的人物包括媽媽的家人、朋友、同事等等，他們或溫和、或熱忱、或敬業、或勇敢，更重要的是，在每一位人物背後，都有一段故事、一份情感或一種人生。比如在 "For My Son" 一詩中，詩人看到兒子對 "true meaning of life" 的精神追求，對兒子的成長倍感欣慰，同時表達了母親深切的愛－"My love for you / is stronger than iron / and softer than feathers, / and if there's a colour to describe it, / it would be the rainbow"。又比如在 "For Cindy" 一詩中，"our angel from heaven, / at the Lily Pond, / in T-shirt and jeans"，一位凡間天使躍然紙上。再比如在 "To Suk Suk, Our Dearest Uncle and Mentor" 一詩中，"If roses grow in Heaven / Lord please pick a bunch for me / and place them in Suk Suk's arms / and tell him that they are from me"，詩句中流露對故人的思念感人至深。

通過每一首詩，我們都可以近距離接觸到一個高品質的人生，而媽媽正是通過她"發現美的眼睛"，把每一位人物的美好品質，都通過優美的語言轉化成朗朗上口的詩句，正如一位巧匠精心打磨每一塊寶石，令其散發出恆久的光輝。正所謂"欲得淨土，當淨其心。隨其心淨，則佛土淨"，我媽媽充滿愛心、天真純粹，總是用愛和善意去對待身邊每一個人，所以她的詩都如山間清泉般清澈而甜美，特別在香港這個紛繁複雜的社會尤為罕有。

除了寫詩，媽媽把自己的生活也過成了詩。她熱愛家庭，給予家人無盡的愛、理解和支持：作為妻子，她幾十年如一日地默默支持丈夫的事業；作為母親，她尊重兒子的每一個選擇；作為婆婆，她從未給我任何說教或刁難，反而從最初認識開始就一直給予我幫助和支持，比如在我剛到香港時，帶我了解香港的習俗文化；在我為宴會打扮而犯難時，幫我一起挑選衣服首飾，並慷慨地送給我各種禮物；因為我不會廣東話，就一直遷就我說普通話；在我們搬家前，為新家的裝修設計出謀劃策等等；她西化、時尚、單純、開明，她與我猶如母女、姐妹、閨密，讓我在"人生路不熟"的香港感到家的溫暖。在這本詩集中，也收錄了她為我寫的一首詩，"And now that you have joined us, / our family is blissfully complete"，她一直把我當做家人／女兒一樣看待，令我非常感動。除了家人之外，媽媽對朋友也十分重視，她的聚會總是高朋滿座、氣氛熱烈，朋友之間都有聊不完的話題。在生活中，她對美有極高的鑒賞力，不僅曾成功經營畫廊，而且在生活細節中也處處體現了對美的追求和品味，無論是每次出門的服飾搭配，還是節日聚會的餐具禮品，無一不讓人讚歎她對美的運用和表現力。她對生活充滿了熱情，政治、經濟、歷史、文化、時事、潮流，都可以成為她的話題，比如詩作"Ode to Knowles Building"記錄了她從大學至今的經歷，其中穿插了重要的政治、經濟、社會事件，與她的感受互相交織，既是一部成長史、也是一段歷史縮影，在詩作的最後，"we soldier on / from shore to shore / from job to job / from crisis to crisis / from year to year / much enlightened / wiser / lighter / and hopefully happier / younger / healthier / and more fulfilled / by the day"，字裏行間充滿了樂觀而勇敢的生活態度。

　　王國維在《人間詞話》裏說:"喜怒哀樂,亦人心中之一境界。故能寫真景物、真感情者,謂之有境界。"從 *My Magic Circle* 詩集裏,我們能感受到真感情、真性情,我們能感受到愛、溫暖和美好,這讓我們從奔波忙碌的生活中得以暫時抽離,讓我們在熙熙攘攘的人間煙火氣裏得以心存詩和遠方,讓我們在白駒過隙的人生裏,得以松花釀酒、春水煎茶。或許,當我們在追逐 Magic Circle 所代表的成功、金錢和名望時,真正的 Magic Circle,原來一直在我們身邊。

　　Anna

Preface

"She said / the other day, / I first saw you forty-one years ago...".

In a state of manifest marvel as she penned these words in a birthday tribute to a friend, our poet was no doubt equally bemused by her subject's vivid recollection, and her professed unconcern for the passage of time.

Thus began a friendship that has spanned the length and breadth of "Ode to Knowles Building", Julia's epic poem expertly interweaving Hong Kong's history with that of the rest of the world through almost as many years, affording that fortunate friend the privilege of witnessing her transition from a talented young lady to the brilliant, multi-faceted poet who now vows to *"free my life with words that I felt, / not with words because they rhymed, / but because they touched my soul..."* ("My Muse, My Light and My Love")

Indeed, Julia only writes what touched her soul, and in so doing takes us with her on a literary journey filled with her thoughts, visions, dreams, joys, and at times, her sorrows. In this the fourth volume of her poetry collection, divided into four sections depicting her beloved family, engaging friends, cherished circles, and finally those sadly departed but fondly remembered, Julia has more than surpassed herself. Although the wonder of her writings remains as arresting and enchanting as ever, her poems are now coloured by a new freedom that comes with an increasing versatility in style and nature: *"I jumped in, / myself forgotten, /… / and my heart open / to her universe..."* ("For Barbara, Crème de la Crème"); a serene charm both beguiling and bewitching: *"her halo tilted sideways, / her robe rumpled /... / and her harp lay mute at her feet..."* ("To Cindy, Our Guardian Angel"); and a tender maturity, all at once soothing and comforting, whilst at times tinged with the bitter-sweet refrains of the inevitable: *"no stone is large enough / to write down / all that / she has offered / to all those / she has touched... / Hers will also be / another life well lived"* ("Celebrating a Life Well lived"), and: *"deep in my heart / I know, / that you will wait in heaven / till God the Almighty / asks you / to bring me home"* ("Stillness of the Dawn"). Such is Julia's mastery in poetic articulation and expression that her often seemingly paradoxical mixing of word intricacies with the simplicity of their delivery invariably leaves the reader transfixed and spellbound, in awe and wonderment.

Warm and humorous, Julia is also fundamentally sensitive, observant, and above all, caring and considerate. She may occasionally be forthright, but her pen is always kind and generous. With this powerful weapon she strikingly describes for us the essence of her "Coronet of Priceless Gems" – the mighty, the effervescent, the God-fearing, the atheist, the warrior, the prima donna, the whimsical, the sadly departed...we live their moments as Julia lived them, feel their sentiments as Julia felt them, and experience their times and tides as Julia experienced them.

Whilst finding her glittering wealth of "priceless gems" mesmerizing, we are equally intoxicated by the sweet fragrance of her "cherished garland." From the sparkling bond in "Ode to My Women Friends", described as: "*like a beautiful firework display... / spectacular colours of the rainbow...*"; to the: "*wild roses... /resembling a halo / of love, friendship and camaraderie*" in "For All My Mediator Friends with Love"; to the unintended but now deeply rooted friendship in "A Tribute to the Orange Party": "*And so the first citric seed was sowed, /unintended, / almost accidentally... / yet the Tree of Friendship took root...*" – the spirit of these idiosyncratic groups comes to life through the vibrant imageries planted by Julia's potent pen; such that just as we admire the glorious fellowship we wish we could share, so do we long for that unique camaraderie to call our very own.

However, Julia ultimately relishes in her roles as wife and mother the utmost. Her dedication and love for her mentor and "lover boy" is most touchingly declared in: "*that you will never go lonely nor be alone, knowing that I will be there for you, for you, for you...*" ("My Dedication to My Most Loving Husband"). In a poem written for her son, Julia the loving mother endearingly recounts the extent of her love: "*my love for you could stretch around the earth countless times and when it is covered, it could extend further to all the galaxies*" ("For My Son"). As for her gentle and graceful daughter-in-law Anna, a young lady of quiet, towering strength beneath a serene demeanour whom Julia once described as akin to a calm harbour amidst a turbulent sea, our poet is equally expansive in her tribute: "*Your beauty lies in the unfading grace / a gentle and quiet spirit / and unflinching courage / and you stole my heart...*" ("For Anna").

Nurtured in love, blessed with grace, and governed by wisdom, it is little wonder that Julia writes with such delicate sensibility and warm compassion – an ability owed in no small part to her not only seeing the virtues and goodness in others, but also appreciating and acknowledging the imperfections and follies within humanity. These, coupled with her remarkable gift of narrative, ensure that the person and her poems are invariably one and the same, and equally exceptional.

Julia's latest collection dazzles with a mix of innocence and depth that intrigues and transcends the senses: "A Circle of Endless Love" warms you, "A Coronet of Priceless Gems" enthrals you, "A Garland of Cherished Friendship" will entrance you, and "A Halo of Gentle Souls" will touch you. To experience these varying emotions, I suggest that you read each poem as your heart takes you, as you abandon yourself to the glorious delight that is Julia's world, and let the enchantment sweep you away somewhere beyond the rainbow's end – to that somewhere by the name of "My Magic Circle".

As I invite you to join me on a kaleidoscopic discovery of the sights and sounds, flavours and textures that are Julia's poems, and perhaps even a little of the talented poet herself, I can assure you that you will not be disappointed.

Trust me, I should know. After all, I have had the privilege of being in the know for 41 years...

Dr. Chiu Ha Ying
Medical Doctor and Paediatrician

Preface

What an amazing treasure trove of jewels in this glittering and colourful collection!

As we turn the pages of this "crown", each "jewel" shines bright, each with its own unique colour, shape, charm and mesmerising qualities.

Particularly moving are the "jewels" in the *Halo of Gentle Souls* section. With her light and unerring touch, Julia remembers and celebrates each of these special gems – there is laughter with the tears, whimsical remembrances tinged with sadness and regret. These are treasures with a warm and mellow glow, soft and abiding hues, forever precious and never to be forgotten.

Then the others, oh, all the others, all her much loved friends and the near and dear ones. How well she knows and understands us and sometimes better than ourselves. With obvious joy and mischief, she picks up each "stone" , lovingly and painstakingly cuts, chisels and polishes, bringing out the facets and the sparkles, turns us to the light, revealing all the radiance in a spectrum and multitude of shades and colours.

She celebrates the fun, the attributes, the good, the foibles and that which is human in us. She peels back the facade, recounts many stories, fleshes out much detail and in the telling brings us all to life.

And Life is after all what this is all about. Running through the pages of this anthology are two central and recurring themes, sometimes muted and barely noticeable but more often dazzlingly bright. They are of course Love and Friendship and through her poetry, Julia pays tribute to these in full measure.

In a way, she speaks for and finds resonance with many of us who could not have survived the trials and vicissitudes of life without these two essential and enduring elements. They are rather like the carbon in diamonds and the aluminum oxide in rubies and sapphires.

Exploring this collection can be an exciting adventure and an enchanting voyage of discovery. The words are from the heart and speak to the heart. In many ways, they are intensely personal and yet the universality is unmistakable. There is much humour in her verse and easy simplicity, cadence and rhythm, an ebb and a flow. The free verse is captivating, descriptive and evocative and it is a delight to explore and get to know each 'jewel'.

There appears to be only one 'jewel' missing – the poetess herself.

But she's there. She's in every word and every line that she's written over the decades and through her work, she shines the brightest of them all.

She is the Jewel in the Crown.

Sharon Evans
Partner,
Foo and Li,
Solicitors and Notaries

Preface

"I write from my heart and use words that flow to me naturally". This sentence epitomizes the magic of Julia's poems. Her poems are more than eloquent expressions of her thoughts and feelings; they are the natural outflow from her heart that she is generously sharing with us. Her poems are the bountiful expressions of her love, her joy, her pain, and her sense of wonder at the people around her and events in the wider world.

In this collection, we see the poet at the summit of her power. This is the poet with the mastery to turn the ordinary into the memorable, and with the confidence in her range and depth to manifest "fire from the heart and passion from the soul". This is an artist whose tools of words give her the freedom to explore how she feels "without pretense nor a conscious effort to impress", and to create.

The story of Julia's journey is wonderfully told in the poem, "My Muse, My Light and My Love". Here Julia traces her evolution as a poet, from the beginning where she followed the strict rules of poetry writing, to her liberation and eventual mastery of free flow from the heart in the open poetic form. The journey has not been easy, but Julia persevered. She was encouraged and mentored by the "lover boy" who breathed life into her poems and her life. This poem is a beautiful boy-meets-girl, boy-wins-girl love story where poetry was the vital cupid. It is a masterpiece among the great poems in the collection.

With her power of words, Julia paints such vivid pictures that we not only think and feel, but also "see", when we read her poems. In the poem "For My Son", Julia describes her love for her only son as "stronger than iron and softer than feathers", and that this love is the colour of rainbow. We can see all that in our mind's eye. We can also see quite clearly the nostalgic scenes of young Julia at university in the early 1970s, described charmingly in the "Ode to Knowles Building". This is a magnificent epic poem covering personal, local and world events spanning three decades, from those idyllic university days to the financial crisis in the 2000s.

Julia's magic circle is inhabited by fascinating people. Through adept portraiture in her poems, she shares with us her family members and friends. Like the best portrait artists, Julia has the amazing ability to convey the essence of a person in a few brush strokes and to bring them to life. Through these poems and pen portraits, we see these people as Julia sees them. We join her in lauding their uniqueness and goodness. We participate in the celebration of their special occasions and daily lives, we smile at their endearing little foibles, and we mourn with her the loss of those who have departed. Julia portrays people in a way that highlights our common humanity, so we feel we know them even without meeting them.

The collection is presented under four sections on the themes of loving family, beacons of light, joy in groups of friends, and in memoriam. The poems vary in length, style and mood – many are joyful, some are funny or sad, and the rest is simply divine. They are all authentic and totally irresistible. Julia engages and draws us into her universe and makes us part of her magic circle. For optimal enjoyment, I suggest you read the whole collection in one gulp, then return to savour each poem one by one.

Barbara Meynert
Business and Thought Leader

Preface

Julia's Poetry IS Julia.

The Magic Circle is a collection of poems on people. Julia once said "I do not write about friends and relatives, you know."

The Magic Circle is also a depiction of Julia's life through her writing journey – her life captured through her eyes of people that matter to her. The poems also brought back memories of the paths we treaded together over time.

Julia was first introduced to me by Carson, my respected alumnus and colleague, as his bride, one evening when they dropped by my home to say hello before attending a dinner function. I remember vividly: there she was, "a gift elegantly wrapped up in a most gentle smile" standing at the door in a white long qipao, just half a step behind Carson – PURE as a child, FRESH as a lily, ELEGANT as a princess and WARM as the summer breeze. Little did I know she was to become one of my best of friends in the days to come.

Julia did not write about friends until about the early 2000s. But *The Magic Circle* recounts individuals and relationships spanning over five decades, from primary and secondary school days at St Mary's and university days at Knowles Building to current year, 2021. Julia is lavishly generous and kind to the people she writes about – these gifts of warmth and appreciation are treasures to keep, priceless to the recipients. She always singles out the outstanding positive attributes of the person to spot light on: insightful, genuine, passionate – "The Ultimate Woman of Class", "She Stole My Heart", "Our understatement" ... Yes, I read about Julia's family and my friends and discovered much about them through her poems. And about myself...

I don't know many poets, but this poet Julia is unequivocally meticulous and precise with her writing. I remember how we 'contended' about the use of specific words – wonder why I was so daring then! But she would have done her research and would justify and defend, with rigour, every word she uses. She puts her soul in her poems.

Over the years, Julia's perception of people and her surroundings further sharpened and refined. The topics she covered broadened as well, from pure portrayal of individuals to incidents and mindset. "My Big Sister", "The Innocent Fish Bone", "A 'High' Ordeal" – brought resonating smiles to my eyes. She even addressed mortality, a topic rather difficult to surmount, not only because of the concept of death in itself, but also the people involved, from the elderly who had led a fulfilled life, to the young who had yet to live. Effortlessly, albeit with an air of melancholy, her words bring solace and consolation to the heavy hearts. "Light a Candle" is delicately touching, and I simply couldn't control the swelling of my tears. This is a journey of metamorphosis for Julia as she unveils herself before her readers.

As Julia lives the life of a poet, transforming to a butterfly as she lifts off, there are however certain attributes that remained unchanged. She is the same Julia: her caring and gentle demeanour; her positive outlook; her compassion toward mankind; the inspiration she emanates; the good she always capitalizes on; and the witty and fun person she is – only multiplied and inspired those that come into contact with her and her poetry. She is still the PURE, FRESH, ELEGANT and WARM Julia first introduced to me, the bride of Carson. And although she has "...no nimble fingers to sew, nor skilful hands to cook, nor the disposition to do housework...", the way she "sees beauty in her eyes, hears music in her ears, says words of love from her lips..." makes this world a better place for those around her, and for everyone.

Julia, on the birth of *The Magic Circle*, we cheer your achievement. We see and sense and feel you in your poetry. Your poetry IS you.

We love you Julia.

Edith Shih
Singing Legal Eagle

Preface

感謝 Julia 將一份又一份珍貴的禮物，也是她收藏着的深情，匯聚成詩集，讓我們一同經歷她從心出發的感動，認識她的摯愛和舊友新知，也透過他們，從更多角度認識這位不一樣的詩人。

正如 Julia 在詩中提到，我們不經意地在朋友介紹下相識。她的真性子和沒有甚麼前設的分享，拉近了彼此的距離。之後，我們無數次再聚，大家無所不談，從日常瑣事，到社會的大小議題，從品味到藝術，我都看見她獨到的觸覺和睿智。我更欣賞她對弱勢社羣的關注，常常探討從不同途徑提供支援。我見證着她全力以赴地帶領社會研究、婦女充權活動和籌款善舉，實踐關愛；她事事盡善盡美，她的動員能力令人佩服！

Julia 重視她遇到的每一個人，她願意聆聽，也觀察入微。她給我的詩，聚焦的盡是美好。原來，我曾分享的點滴、曾經歷的掙扎和不容易，她都牢記，她的描繪是那麼深刻，也充滿鼓勵，讓我感動！在這本以 "人物" 為主軸的詩集，我看見更多她筆下美麗的故事，和她對生命的禮讚！

我很喜歡詩集中的 "My Muse, My Light and My Love"，詩中説詩人和她另一半相知共賞、互相啟發、彼此激勵，在 "詩潮" 中並肩前航。因為他的支持，Julia 更堅毅努力，她駕馭一切的規律框架，"我手寫我心" 地，將她對人和事的熱忱，徐徐活畫於詩篇中，讓文字的心和意躍動，連繫她和詩中人，也牽繫讀詩的人。

Julia 寫詩，也寫情、寫理、寫智慧。詩篇中，我看見詩人以細膩筆觸，豐富、優雅又言猶不盡的辭彙，刻畫每一位詩中人的獨特，綻放他們瑰麗的光譜，互相輝映；我也深深感受到詩人對摯愛家人的感激、對朋友的欣賞、對故人的惦記；字裏行間，我窺見她對生命的哲思，對社會及家國的情愫。這些詩也引導我有許多反思，為我帶來啟迪。

縱有言情説理，Julia 的詩從來不造作、不説教，而是載着令人會心微笑的趣味。在她筆下，生活的小片段都展現姿趣。其中兩首詩："The Innocent Fish Bone" 及 "A 'High' Ordeal"，更帶我進入她曾度過的憂驚，我一面讀詩，一面急不及待追看究竟，我為結局感恩，也被她的幽默懾住。

　　詩集中，有不少引發我共鳴的詩章，讓我回味幾許的人和事，和久違了的思緒。詩人的 "Ode to St Mary's" 及 "Ode to Knowles Building"，更讓我穿越時空，置身當年的校園，回味少年十五二十時的精彩。驀然地，數十年的略影、社會的蛻變，在腦海中閃動，感激也感喟！

　　正如它的名字，這本詩集記載着 Julia 那叫人稱奇的"圓繫"，而她，也只有她，就是那永不動搖的軸心，連繫着這一切的人和事，繼續發掘和發揮當中的璀璨，叫一切顯得更圓滿。

　　我誠意推薦這本詩集和裏面每一首詩，也鼓勵讀者細嚼當中的詩意深情，大家透過詩，圓繫、圓滿。

Amarantha Yip
Chief Director,
Hong Kong Family Welfare Society

Preface

一直以為，送你一首詩，是古人交心、或者是情侶相悅的表達，沒想到，在文字幾乎被丟棄的年代，我會收到如此獨特的禮物 – 一首詩。

認識 Julia 只是這幾年的事，有些人，天天相遇都未必頻道相同；但有些人，未曾見面已經心靈相通。我想，我跟 Julia 就是這種朋友。

我們不是在行人路上碰面，我們是在思想的天空中交流，好多事情，我們都想到同一道上，彼此擊掌呼應，德不孤，必有鄰。

Julia 有種洞察人心的特異功能，她能看穿一個人，然後化成一首詩，觸動人心。看穿人還容易，賦成詩卻極難，尤其在今日影像橫行的世代，大家愛看懶人包、愛做標題黨，文字漸被遺忘，更何況掌握它的韻味。

其實我從小就愛詩詞歌賦，唸中大中文系的日子，專挑唐詩宋詞、古典戲曲、蘇東坡辛棄疾的研究來選修。年輕時愛參加寫作比賽，年年都送稿去「青年文學獎」，唯一一次有收獲的，就是拿了首詩去參賽，那年，背着背包在絲綢之路走了一個月，除了賦詩，我不能用其他形式來表達深情，於是，在天山腳下、在香妃墓前，我記下青春的感悟。回來，把詩作拿去參賽，竟然得獎，那時我想，我要做個詩人了。

然後，唸書、考試、工作、結婚、生孩子……漸漸，忘了詩。

這幾年，拿着筆捍打仗，寫的是評論，離詩人的優雅，漸行漸遠，有點逼上梁山之感，那明明不是我的路。

直至那一天，收到 Julia 送我的一首詩，如電擊般怦動，詩興都回來了。為有犧牲多壯志，敢叫日月換新天，我們能為時代出一分力，另闢新徑又何妨？

Julia 的細膩讓我感動，我沒她筆下那麼出色，但她的詩卻讓我豁然頓悟，從此，我不僅要剖析世情，更重要是感動人心。

詩之偉大，是它能觸及靈魂，我習詩多年，卻在俗世洪流中忘了初衷，感謝 Julia 的詩作，您讓我重新思考文字的意義。

屈穎妍
資深傳媒人

Acknowledgements

I am greatly indebted to many of my wonderful friends who have made the publication of my fourth poetry book possible.

My Magic Circle comprises poems I have written for many of my friends who have left imprints on my heart. I want to say thank you to each and every one of you who have entered and touched my life in your own unique ways, enriching mine with such colour and light, sound and music, art and magic, and for that I am eternally grateful.

I owe my deepest gratitude to my five amazing friends, Chiu Ha Ying, Sharon Evans, Barbara Meynert, Edith Shih and Amarantha Yip, each of such eminent distinction in their professional fields, for writing the Preface for my collection, having to read all my poems when they are so time pressed. I am most grateful for their kind and generous endorsement.

I like to thank Betty Wong, my editor at The Commercial Press who has worked on every of my poems with such energy and indeed professionalism, rendering the final product with a style that is both elegant and refined. The poems have been arranged in chronological rather than alphabetical order and the date of when it's written is at the end of each poem, a habit I have started since I embarked on my poetry journey.

I am particularly indebted to Jimmy Fong, an award-winning graphic designer friend of mine who gave me the magic book cover design and selected a unique font for the text that is poetry itself. I am grateful for his professionalism, his dedication and, above all, his generosity. I am also most indebted to Ruth Lau, my very artistically gifted friend, who has designed my section breakers with her vivid silhouette illustrations using only a thin painterly pen.

I am greatly blessed by the encouragement and support of many of my friends who tirelessly responded to every poem sent to them as soon as I finished one. Their warm reception has been as reassuring as it is encouraging.

There are many other dear friends of mine whom I have started writing poems for at the moment, and they will be featured in my next collection, hopefully before too long.

Finally I wish to pay tribute to my three very best soulmates, my husband

Carson, my son Ian and my daughter-in-law Anna who have been my constant source of joy, inspiration and support.

May I take this opportunity to thank you all for this journey of reconnection with friends whom I have written on so much earlier and engagement of those I am inspired to pay tribute to during the past two years. It is to me a journey of the heart, and one that I shall keep and cherish for the rest of my life.

A Circle of Endless Love

My Muse, My Light and My Love

St Paul saw the light on the road to Damascus,
Archimedes while soaking in the bath,
my moment of illumination
was yet to come....
My first attempt,
still in my teens,
an eight liner,
closely imitating those of Byron
with a little of Keats thrown in,
a gloomy one,
about life and death,
and read only to a girl I hardly knew –
"it rhymes," that's all she said,
mercilessly pointing out
the only feature it has.
She's right –
that's all there's to it,
really,
entirely without merit
or
promise....

Then he came,
this lover boy,
showered me with poems
so tender and sweet,
so sensitive and passionate,
so sentimental and delicate,
so ardent and moving,
so poignant and intense,
and yet so powerful
they ached my heart.
He sent me one every week,
his "Monday Ritual",
our little secret code,

complete with a floral arrangement,
that would come
at around 3 pm,
non-stop,
uninterrupted
for over two and a half years
with one even on our wedding day
which did not fall on a Monday...

I was so inspired
that I wanted to write too,
and indeed I did,
no less than fifty,
all in those hot summer months of 1981....
I locked myself up in that little box,
of meters and forms,
imageries and rhymes,
endlessly wrestling with words and images;
My anthologies of poetry I piled up,
Roget's Thesaurus I put close by,
from mosquito bites to mah jong play
I churned out one in less than a day,
"you are so witty," they all said.
But inside me I knew,
it's just not there,
it's just not there yet....
They were funny,
lyrical like songs
but poetry is more than words and sounds,
it's fire from the heart
and passion from the soul...

Rhymes restrain the inner flow,
dull the senses,
confining,

stifling,
leaving me stiff....
I yearned to fly away from that box,
to free my life with words that I felt,
not with words because they rhymed,
but because they touched my soul;
I wanted to express my passion
with words
that have a heart of their own.
In desperation I cried –
no more pretentious words,
no more forms,
no more structures...
How I did try,
again and again,
I tried,
but without that box,
no words came...

My lover boy came,
he came to my rescue,
unfastening my little box
with his key of love.
He's my beacon of light
and would encourage me
to keep up writing everyday
like a daily routine,
and do so
without expectation nor exasperation.
"Persevere is the only answer"
so the poet warned.
Read as much as possible
to widen your horizons,
stay intellectually curious
and interested in everything around you,

but above all else,
get yourself actively involved in the outside world
that's at once as ugly as it's beautiful,
as complex as it's simple,
as tumultuous as it's agreeable,
as dynamic as it's stable
and it's up to us
to explore and discover our true selves;
and then,
gradually but surely,
the subject will find you.
The best poems get written,
not by going in the front door of the subject,
but round the back
or through the window
when one least expects it.
And as I opened my heart
and started to learn to appreciate
all the little things around me,
I was able to find,
to my huge surprise,
magic right under my nose
and that simple ordinary things
could be so extraordinary,
fascinating,
even irreplaceable and profound,
just kind of marvellous,
almost breathtaking.
My journey I started,
hesitantly,
diffidently,
one step at a time,
with endless,
unceasing,
positive and constructive encouragement and support

from the poet himself,
with so much revision and editing
in the beginning
where even one notion,
one idea,
a single phrase
or even a word or two
added or edited out
would change the tone or aesthetics,
the tempo or flow,
the grace or charm
of the piece,
and for that I am eternally grateful.
It's like he's dotting
'the eye' on the dragon
for me
and I shan't know
what I could achieve without him.
For I will never forget
what he has inspired in me,
not just by the poems
he has written for me,
but the way he has improved mine
through the years.
I no longer feel the need
to use exclusively abstract words
or flowery language
to write complex verse
to convey deeper message.
I write from my heart
and use words that flow to me naturally
and give myself the freedom
to explore exactly how I feel,
without pretence
nor a conscious effort to impress.

I write anything that touches me,
anything from the ridiculous to the sublime,
from the tragic to larger than life,
from art to fashion,
from idle thoughts to contemplation,
from statesmen to heroes
to my own family and good friends.
My poet
has breathed life
into my poems
and my life,
as we sail through ours together,
with joy and in wonder,
come rain or shine,
through thick and thin,
in sickness and in health.

February 2021

For Anna

As my son grew to manhood,
I prayed
and how hard I prayed,
that he would one day
find someone
to share his life with:
someone who loves him for who he is,
someone who's not only gentle and kind,
but sensitive, considerate and empathetic
that he feels comfortable
to share with her
his aspirations and hopes,
his fears and worries,
his challenges and frustrations;
someone he could trust with
all the secrets of his heart,
his triumphs and failures,
his attainment and disappointments,
his felicities and miseries
that life inevitably brings,
that two hearts
could rejoice and weep together
as they go through
their life-long journey,
supporting and blessing each other,
over the long, long years,
with joy and in wonder,
in health and with purpose.

And you came
and my prayer for him was answered.

I still remember the time
when you and I first met –
and how struck I was

by the innocence of your face,
the face of a child,
no more than 18,
a mere college girl...
There's this aura of purity about you,
so refreshing and unaffecting,
and yet so gentle and sweet...
Tall and slim you stood,
smiling shyly,
and when I beckoned you over,
you came
ever so softly,
like a whiff of a summer leaf,
responding, acknowledging,
nodding, rejoining,
ever so amiably
and in a manner
that is as mild as it is gracious,
and expressed
so eloquently and appropriately,
that I am to learn only much later,
that you are not the top student in school
nor a master degree scholarship holder
for internship at Baker
for nothing.

And you stole my heart,
then and there.

Your beauty
lies in the unfading grace
that comes from inside you:
of a gentle and quiet spirit,
an indomitable will power to excel,
and an unflinching courage

to look for the star.
You are gentle, soft and kind
but if they are interpreted for weakness,
one is very much mistaken.
Yes, you are mild, soft and kind,
but I have not seen
a more resolute determination
nor fearless self-awareness,
nor unbeatable conviction
to dare venture
into the unknown
that you might build a kingdom
after your heart.
In leaving your hometown
and coming over to Hong Kong,
a city with such cultural differences
from where you were born
and brought up in,
with neither relatives nor friends close by,
you did the bravest thing
that would unnerve any woman
of lesser courage and fortitude.

Young as you were then,
you know where your heart laid.

So here's a woman
who works more than she says,
and thinks more than she speaks.
A doer and not a talker,
she gets things done
with minimum fuss and fluster.
She finds her strength
in her inner core
and nothing could dim the light

which shines from within her.
I still remember how you
helped me translate
my English script into elegant Chinese;
put my ideas of a speech
into an effective presentation;
improve my Putonghua pronunciation,
to make sure every single word
is delivered correctly,
order online products,
basically doing what a daughter would do
for her tech-ignorant mother.
But what I am most grateful
is your unconditional love for my son,
the perfect couple
that must be a gift from Heaven
that a mother,
even the best in the world,
could just be blessed that much...

And now that you have joined us,
our family is blissfully complete.

Much love,
Mum
January 2021

For My Son

It seemed but yesterday,
when I sat next to you,
your young curled-up fingers
on the piano,
playing Clair de Lune,
for me,
your tone-deaf mother,
the only piece she knew how to enjoy then,
and now still...
and suddenly
with a blink of an eye,
my little boy is all grown up,
having studied in different countries,
earned a few degrees,
a lawyer initially
and now a banker,
just like his father,
happily married,
and about to grow a family of his own...

For you have taken time to understand yourself.
You started early,
much earlier than most,
when your peers
were still playing football
or chasing girls.
You wanted to search for
the true meaning of life,
and the reasons
why there are so much
pain and suffering,
injustice and inequality,
or oppression and abuses,
in almost every part of the world.
And you found some of the answers

in Buddhism,
a spiritual tradition
that focuses on personal spiritual development
and the attainment
of a deep insight
into the true nature of life.
And now as a practicing Buddhist,
you believe that life
is both endless
and subject
to impermanence, suffering and uncertainty,
and the only way to bring us
secure and lasting happiness
is internal transformation
through enlightenment
by following a spiritual path...
And so my dear son,
young as you are,
you have reached a state
when I could no longer be your light.
But please let me continue to be
your soulmate and companion –
do share with me
your thoughts and dreams,
your worries and anticipations –
let's always make a special time
to continue a habit
we have started
long before you could speak,
sharing,
total sharing,
absolute sharing,
your joy and sorrow,
your achievements and failures,
your ups and downs,

your triumphs and defeats,
your glory and scars.
Yes, I remember all those scars
we went through together,
and how you triumphed over them.
No one has yet been spared of scars,
and we can't make them disappear
but we can change the way we see them.
I am so impressed that
you see them as the tattoo
of a triumph to be proud of,
a sign of strength and not pain
only the courageous and the positive
could manage.
I see the best parts of myself reflected in you,
but you have made them even better.
So just let me tell you
how proud I am
of you and your achievement.
I am so pleased that
you have found your path,
and I just want you
to be true to yourself
and that I will stand by your side
on whatever you decide.
There are no words to express
how much you mean to me,
for a son like you,
the ultimate definition of a son
in every way,
is a blessing from Heavens.
My love for you
could stretch around the earth
countless times
and when it is covered,

it could extend further to all the galaxies.
My love for you
is stronger than iron
and softer than feathers,
and if there's a colour to describe it,
it would be the rainbow.
My son,
you are the very best and I love you so.
I am thankful for every moment
that we are together,
and thank you
for making me feel so blessed.

Much love,
Mum
January 2021

For My Most Loving Husband

(Inspired by Sylvia Plath)

I took a deep breath
and listened to the old brag of my heart:
I am, I am, I am...
I might not have nimble fingers to sew,
nor skilful hands to cook,
nor the disposition to do housework,
like a good wife a husband expects her to be,
yet I see beauty in my eyes,
hear music in my ears,
say words of love from my lips...
And grow flowers in my heart
to weave a garland of love for you
that you will never go lonely nor be alone,
knowing that I will be there
for you, for you, for you...

November 2018

Birthdays for My Two Men

He just passed his birthday,
my husband,
marked but uncelebrated,
as he was somewhere in the sky
only to arrive home
after midnight
when it's officially past.
Not that it matters really –
birthdays have become
so small these days,
we have,
for quite many years by now,
allowed them
to pass quietly by,
unintentionally mostly –
a candle-lit dinner for two,
a banquet with family or good friends perhaps,
a movie at home,
a short stay in Tokyo or Shanghai,
a present or something,
a simple scarf or a diamond ring,
a word or two,
a poem,
anything goes,
whatever the mood takes us,
nothing elaborately planned,
and not always on the exact day,
not when there are
more important
or trivial
social or whatever duties
to oblige...

For us,
birthdays have become smaller

as we grow older –
when we were little,
every birthday was a big one.
It took us such patient waiting
for it to come
and when it did,
we were so excited
we could hardly sleep the night before:
a new dress,
a pair of matching shoes,
so many presents,
so many friends;
no home work to rush,
no revision to do;
dad took a day off to join the party,
and mum all smiling and happy;
then counting the candles,
fervently making a wish
before blowing them;
dishing out the cakes,
the one with the largest strawberry
for our best friend;
the feverish opening of the presents,
one after another,
with no one being allowed to help...
That was our special day
and we were the special child,
so thrilled,
so contented,
so carefree...

Ten felt very, very big,
those two digits,
one so straight and grown-up,
the other so round and promising;

And then 13,
just into the teens
when childhood was but a memory;
16 was sweet,
and 18 was freedom
when we fell in love.
At 21,
we became adults,
with the whole world on our palms...
But then at some point,
without us knowing how and why,
it just dawned on us
that 28
was not too different from 27,
or 46 from 45.
Your age in years
seems to detach
from your age in experience.
You get fired at 32 and feel 12,
vulnerable and teary;
you get a standing ovation for your speech
and feel 12 again,
vulnerable and teary.
The health culture
also conspires to add to our confusion:
now that 50 is the new 40 is the new 30,
and when you have a child like I do,
he is going to mess up your life cycle even more...

Each of his early birthdays
loomed so large
that ours simply receded.
When our son was ten,
he quietly asked me,
"must we have the clown again, mummy?

I don't think my friends think he is cool!"
We have had clowns
for ten years
and if that's enough for him
that's enough for me.
I replied seriously to his serious question:
"no, not if you don't want to!"
And now that he is reading his third degree,
I too have allowed
his birthdays
to become small
as requested.
And now when I do have a birthday cake
in my honour
and watch the candles flicker,
I do not count the number
but the many past birthday wishes
that have come true for me,
feeling so blessed and privileged.
For years now I have made birthday wishes
not for myself
nor my loved ones
but the less fortunate
who need the wings of angels
to lift them up
every now and then...

Let's make everyday
a birthday for someone
no matter how small it is!

May 2009

For My Dearest Son

This is a difficult summer,
a summer without you –
Shall I take up this course
in Shanghai for three months
learning how to do business in China
which also offers internship at the last month?
My heart sank
but I readily agreed.
Whatever good for you is good for me...
And in Shanghai,
you seemed busier than ever with your classes,
and when you started working,
you had to wake up so early
and come back to the dorm so late,
I wished you were here with us.
August came and you said:
come over with dad
so we could spend your birthday together
just like what we have done every year.
No, you have not forgotten my birthday,
and like I'll never forget yours,
it seems you will never forget mine either...

It was the most tearful goodbye,
dry without tears,
swallowed for each other,
you were but 13 then,
away for the first time
so faraway,
all on your own –
"I'll fly over for all your birthdays,"
I said,
something I must have said a thousand times –
for I was going to cry,
any minute soon,

and I needed to say something,
anything...
my heart was aching,
my fortitude crumbling,
falling to pieces...
"Good mum",
you said,
"we are lucky, aren't we,
my birthday falls within the mid-term break."
The English are so organized,
the school plans out a five-year schedule for you.
Amazing, aren't they!
Yes,
you were amazing,
so young and yet so mature,
you tried making it easier for me
when you were gnawing inside,
I knew it,
I knew it,
I just knew it...

We had dinner
at this fancy Italian restaurant,
you were so full of music,
we kept talking about
the musical we just enjoyed,
how you whispered the lyrics
in between the meals,
and after you blew the candles on the birthday cake,
you pushed your chair next to mine,
and started singing softly,
all the wonderful songs,
one after the other,
one after the other,
like you didn't want to stop...

I remember
how you sat,
how you sang,
how your body swayed to the music,
how you looked at me,
how the people looked at us...
The picture is in my heart.
You were 14.

Years swept by,
everything happened
but yesterday
and suddenly you are all grown up,
still my best friend,
my soulmate,
my confidante,
my love,
forever and ever...

Mum

August 2006

Happy 21st Birthday, Darling

"Shall we go now?"
my son asked me.
Now?
With me?
On his first morning
of summer break
and not with his girlfriend?
Surprised
but deeply moved,
I gently said,
"yes, let's go."
For hours we stayed
at Tom Lee,
he trying
every model,
every key,
his fingers
feeling the touch of each,
his eyes
devouring the beauty of ebony,
his ears
capturing the varying nuances
in timbre and tone...
and I,
not knowing Yamaha from Steinway,
stood by,
waiting,
listening to his findings –
music is not my poetry,
he is...
When he was five or six,
I wanted to buy him a piano,
he just passed grade 2.
"Don't buy it for me,
I hate playing the piano" –

in such a tone of finality
a mother knows
there is no compromise.
In desperation
we reached a pact:
he did not have to practice
if he didn't want to,
no more forced practice,
no more exam –
just sit for the weekly 1-hour lesson –
that's all I asked.
For years
I reluctantly
kept my promise,
he most happily
kept his.
The piano sat idle,
untouched,
uncherished,
unwanted,
except for that hour in the week...
Then the Phantom of the Opera came,
and mesmerized Hong Kong...
One fine day as I came home from work,
I heard the most wondrous music
flowing out from his room...
my piano-phobic
was practicing
on his much neglected piano,
his fingers flying,
his eyes reading the scores,
his body swaying a little,
the metronome went
tit tat, tit tat, tit tat...
I stood outside his room,

listening quietly
until he stopped...

He has never stopped since.

Happy 21st birthday, darling,
you deserve
the grandest piano
love could buy
and
thank you
for letting me
come choose
with you.

May 2006

A Coronet of Priceless Gems

For Winnie
The Enlightened One

"Why don't you join us for a business lunch",
so my husband said as he left for office.
What a rare suggestion,
but I knew,
I knew then and there,
that I would be meeting someone very special.
And he's right,
like he is,
most of the time...
My husband did most of the talking,
the president she brought asked many questions,
she raised a few,
and I sat there,
said nothing,
listening,
and smiling for no reason,
just feeling happy.
Anyone can make you happy
by doing something special,
but only someone special
can make you happy without doing anything.
And she's that special someone
who came into my life
by divine providence,
enriching my life,
as I get to know more of and learn from her.

It's her demeanour,
the way she carried herself,
that first struck me –
so peaceful and serene,
so modest and unassuming,
so genteel and at ease with herself,
and all the more extraordinary,
for me to learn later

that she's a businesswoman of huge success,
and a philanthropist held in high esteem.
She invited me to visit her office after lunch
and I was unprepared
for the precious collections
I was introduced to,
one after another,
each a mini museum on its own.
I was mostly touched by her Buddhist collection,
my son being a devout Buddhist himself:
There are Buddhist statues and sculptures,
plaques, panels and prayer beads
in marble or jade,
in silver, bronze, or copper,
in coral, bone or wood,
in lacquer or clay;
and manuscripts and paintings
and many others...
She's no scholar in Buddhism
and so she very modestly told me,
but her deep connection
with her Buddha nature
is plain for us to see,
a self nature which exists in all sentient beings,
lost in most of us
as we go through the challenges
of such a multitude
of material and sensory temptations every day.
She told me how she met her guru.
When she was younger,
a colleague took her
to Danxia Mountain in Guangzhou
and they met an eminent monk in the temple there.
His Holiness told them to walk around
and come back for dinner.

As they strolled along leisurely,
her colleague asked her
something most unexpectedly:
would His Holiness accept you as his pupil?
She laughed and said:
if he could make me 'always happy'
I would be happy to follow him.
When they returned to the temple,
His Holiness said he liked to take her as his 'tu di'
and gave her a folding fan
with 4 characters already written on it:
常興愛徒 –
Always Happy dear tu di –
she looked at her given name,
too stunned and overwhelmed for words,
and without realizing what she's doing,
her knees went down
as she bowed before her master.
One might search a lifetime for a guru
and still fail,
but hers came to her just like that,
yes, just like that,
to the envy of all eager seekers.
And that started her lifelong journey
of charity and compassion.
She was once in Shenyang
selling shops
of her newly developed shopping mall.
A lama came over to her
and said they needed
to build a Buddhist College in Tibet.
And with not the slightest hesitation
she gave him the money inside a box
she just collected from her sale of shops,
knowing that more would graduate

to spread the Buddhist truths and virtues to the world.
Then just as she has almost forgotten about it,
she received news
that the campus was completed
and she's invited to officiate the grand opening.
She has since continued to give
without expectation of any return,
and she's rewarded abundantly
with joy and happiness.
And she has built over the years
more than 40 temples.
She inspires others by her deeds
and not her words
which she utters little.
She's kind and loving,
patient and understanding,
insightful and open-minded
and she's so completely oblivious of self
for she understands the impermanence of life
and any clinging to or being obsessed
with the delusional self
will bring about suffering.
Her guru has awakened her
to the nature of human suffering
and she makes it her mission
to help others find freedom from such suffering.
She leads by example rather than control,
and people follow her
because of who she is
and what she stands for.
We all want to be more like her.

And so Winnie,
my enlightened friend,
this is my little tribute to you

to tell you
how grateful I am
that you have come into
and touched my life,
and the lives of many.
You are a spiritual messenger,
and I am waiting for my own awakening.
I know it would not come
like a lightning bolt
or a sudden flash of sparks.
It will come in unexpected ways
and I will make myself available
and trust that it will find me
eventually...
while I will,
like you,
continue to be of service
to our community,
giving myself
in service to our common enlightenment,
and passing the bright flame
of hope and love
from one warm heart to another.

July 2021

I just came across Ovid's famous quote
the other day –
"That you may be beloved, be amiable"
and I thought of her,
the young lady I met
when we were in our 20s,
in our first flush of youth,
she at the Graphic Design Department
and I the Editorial.
I still vividly remember
the way she looked and carried herself
when I first saw her.
There's this rare grace in her manners
for someone so young –
so genteel and soft,
so refined and gracious;
and her face,
so exquisite with such flawless porcelain skin,
a modern princess came alive
in my young mind,
despite her shocking mass of tangled curls,
flying out in all directions
in full volume,
and her oversized shirt complete with ultra-tight jeans,
her signature fashion style
so I soon learned...
She was my modern live Barbie Doll
my first and my last.
But it's her goodness of heart that has captured me:
I could not remember how we became close
as we started spending
more and more of our lunch time together
either in the office
or the restaurants nearby;
but there's a poignant moment

that touched my heart
when she did something extraordinary for her father,
now already in Heaven,
when I knew then
that I would like to have her as my special friend,
and indeed our friendship
grew and blossomed over the years
until this very day
when we still manage to have our monthly lunch
that started in recent years,
even during the darkest hours
of violent riots and frightful Corona pandemic.

She's indeed the gentlest person I know
and one of the most loyal and trustworthy.
She's always polite and kind,
truly respecting individuals
and treating 'kings and paupers' alike.
Yes the tone she takes or the way she talks
to those in high society
is the same she uses on her helpers or driver.
She would readily give a helping hand to a stranger seeking it
and not with the slightest hesitation.
She is always happy to go the extra mile,
whether it's staying behind
to help clean up after an event
or spending her own time
to ensure things get done properly.
She has been our pro-bono PR Manager for decades
whenever our art gallery
opened an art exhibition,
even at a day's notice
when me and my partner,
in the hectic struggle
to get everything well organized,

forgot to inform her of the date.
Her generosity of spirit knows no bounds.
She would arrive,
elegantly bejeweled and in her finest,
just to give us face,
sometimes even earlier than us,
go around entertaining our guests,
and never forgetting
to bring her well-off art-lover friends
many of whom have become our most valued clients.
She's just happy to see us happy
and would decline
even to join our celebratory dinner
believing that we should
only spend time and efforts on potential buyers.
There's never the need to say thank you,
she would tell us every time.
A truly generous temperament with an indulgent heart.
So one knows why she has so many friends
and is someone people like to be around with:
for she's always so kind,
never finding faults in others,
so non-judgmental,
and always finding the best in a person.
A true friend she is.

And a true wife and mother too.
She loves her husband and two sons
and respect them as individuals
with flaws and all.
She is their best friend.
She's supportive of them
whether it be a career,
a hobby or anything else
they are interested in or wishing to pursue.

She's there for them
even if it means
having to travel more than six months in a year;
or having to learn how to golf for her husband
so she could enjoy an activity
that both of them can do together;
or attend cooking lessons
so that she could make a feast for them
when they return home.
And what a professional golfer and celebrated chef
she is now.
I don't know much about how she learns how to golf
but I do know
how she did not even know how to boil a kettle of water
when I first knew her
when her favourite recipe
was either 'order in' or 'eat out',
and when she had rarely entered
her large kitchen in her huge maiden mansion estate
except for a glass of wine or juice.
And now her greatest joy
is going to the wet market every day
and cooks for his three men
who turn hungry boys
the minute they are at home
waiting for dinner.
She loves to cook for their friends
as her husband is a partying animal.
I have seen her in action
and I loved the way
she decorates and makes each dish
so gorgeous and appetizing.
It's like she's an artist
showcasing her creativity in each plate
and making every guest

feeling so special and loved.
Her secret ingredient is no doubt
love, love and love.

Now in case you may wonder who this lady is
for she may seem too good to be true.
Please show this little tribute
to your three men
and they will tell you who she is.
My dear friend,
let me take this opportunity
to tell you how wonderful you are
and how blessed I am
to have a lifelong friend as fine as you.

June 2021

For Eunice
A Devout Atheist in Search of Spirituality

It's a surrealist journey
of amazing discovery,
even for a fellow thinker like myself,
as I went beyond
her flawless,
translucent porcelain face,
complete with bright sparkling eyes
and full luscious lips,
and a near-perfect physique
twenty years younger than her age,
to the depth of her mental prowess
and cognitive capacity,
in full display
after an in-depth sharing
and exchange of thoughts and ideas.
Everything she does –
be it for physical appeal or fitness,
intellectual pursuits,
or leisure interests –
she gives 200% efforts
to achieve the perfection
she desires.

The pilgrimage began
as the door opened to us
to a huge home of laser lights,
rough walls and coarse surfaces,
panelled and spiked ceiling,
cabinets with an assortment of fossils,
relics and pre-historical remains,
fish tanks,
bookshelves of serious hardcovers,
and a rhapsody of colours
which includes
not only a collection

of electric blue,
Burgundy maroon,
fluorescent green,
cardinal red
and mulberry purple sofas and armchairs
but a vivid red Steinway grand piano
sitting at the very centre of the living area
all for our much jostled mind
to visually create an order
within the intended chaos...

And indeed an order there is
not easily understood until being explained
and explain she did:
The main theme of her home décor is
all about the origin of our universe
and the evolution of life by natural selection.
The powder room says it all:
it has been transformed into the universe,
and the numerous laser lights
represent the planets and stars,
while the spectacular laser effects
mimic supernova, nebula, and other celestial events.
The spiked ceiling
represents the wave function,
a quantum theory,
which explains the nature of the particles
that make up matter
and the forces with which they interact.
The coarse surfaces
simulate our young earth,
while the collection of fossils and remains
the footprints of evolution.

For someone who took such immense efforts
to understand the evolution of life,
and has her dining table specially made
to reflect the 4.54 billion years
of the history of the earth
in a 24-hour timeline;
and an ardent student of Dawkins and Krauss,
who are advocates
for public understanding of science,
and public policy based on
sound empirical data and scientific concepts,
it's no wonder
that atheism is her religion.
She contends that a supernatural creator
almost certainly does not exist,
and that belief in a personal god qualifies as delusion,
which she defines as a persistent false belief
held in the face of strong contradictory evidence.
She is truly a devout atheist
who disbelieves in the existence
of God or gods,
a belief she practices religiously.

So how does she,
an atheist,
find meaning in life?
While most find
financial gains and profits,
creative pursuits,
travel,
or leisure activities
as meaningful,
she finds it
in spirituality,
something she continues

to keep searching
in earnest.
The understanding of the wonder and beauty
of how the universe
or how life has evolved,
is to her the utmost spirituality,
surpassing all other religious myths.
She feels the need
and more than a responsibility
for her very own self,
her loved ones and indeed everyone,
to appreciate our brief moment 'under the sun'.
She makes it her mission
to help others to understand it:
to open their eyes,
their minds,
and their hearts,
through science.
As her teacher Krauss has said:
a universe without purpose
should neither depress us
nor suggest that our lives are purposeless.
Through an awe-inspiring cosmic history
we find ourselves on this remote planet,
in a remote corner of this universe,
endowed with intelligence and self-awareness,
and instead of despairing,
we should humbly rejoice
in making the most of these gifts,
and celebrate our brief moment in the sun.
It's such a pleasure knowing *you*,
my atheist friend,
and thank you for telling me
that I exist
because the stars were kind enough to explode

and while I now exist,
no matter how brief,
I am going to appreciate and enjoy this precious gift
to make it not only heavenly
but meaningful.

April 2021

A Birthday Tribute to Ada
My God-fearing Friend

"Charm is deceitful, and beauty is vain,
but a woman who fears the LORD is to be praised."
Proverbs 31:30

She is God-fearing,
this my dear friend:
it's her attitude of respect,
a response of reference and wonder
to God the Almighty,
and a commitment
to stay completely away from sin
as a child of God.
I first met her
in a bible study class at her home
some fifteen years ago.
Her goodness of heart
was etched all over her face,
and in the warm and caring way
she made me,
the new comer,
feel so welcome,
as she gently made sure
that I was comfortable,
being introduced to everyone,
had a good dinner,
and ready for class.
This her engaging and inviting smile,
her goodness of heart,
and her generosity of spirit
have gone a long way
into my heart
and continued to this very day.

She is God-fearing,
this my dear friend:

she lives her life
devoted to the love of God
and in obedience to His Commandments.
She sees things of beauty in all that she sees,
when a weed to most is a flower to her,
for beauty goes with her wherever she goes.
She sees the good in every person.
She's never judgmental
as she is aware of her own imperfections.
Instead of walking away from someone
who has messed up her life,
she chooses to believe in her potential
to change for the better.
Instead of avoiding the person,
she makes an effort to reach out
and encourage her
not just with words but solid help.
She takes no part in gossiping
as she knows gossip is toxic
and would shut her ears to those who do,
and keep quiet
if there's nothing good to say about someone.
She is totally trustworthy
and we all feel we could confide in her
knowing that whatever
is said will be kept inside her forever.

She is God-fearing
this my dear friend:
she is receptive to God's plans
and has faith that they are the best for her.
She's most considerate
of the feelings of those around her
and has great compassion
of those in need.

She's always there to lend an ear
to listen to the pain and suffering of others
and a shoulder for them to cry on,
while she puts hers aside
and let God come to her rescue,
for she relies on Him
particularly through hard times.
And I have never once heard her complain
about the ills of life she has to face
which have not been any lighter than most of ours,
and that she has always endured them
with stoic fortitude,
patiently and quietly,
for her faith is unwavering,
and she believes that God has given her strength
enough to weather every storm,
and that the sun
would eventually come out of the clouds –
and it has
every time.

She is God-fearing,
this my dear friend:
she has lived to show me
that "the prayer of a righteous person
has great power as it is working."
I have often travelled with her
and had shared a room with her a few times.
And no matter
how early we had to leave the hotel room
or how late we came back,
she would never forget
her morning or night prayers
or prayers before every meal.
She's shy and quiet

but when she prays for you,
her words of love so tenderly said
would renew you,
for the soothing comfort and hope
they make you feel.
It's like God has empowered her to bless us.
And she would pray
to comfort those who are troubled,
or sick or worried,
or sinking.
She would pray for our spouses and children
and for those seeking help in times of crisis.
And her prayers are the best gifts
a person could receive from her.

She is God-fearing
this my dear friend:
she lives a generous life
which is a continuous act of love,
solidarity and dedication to others.
She lives with her hand open:
open to hold the hand
of a sobbing child or handicapped;
open to hold wrinkled fingers twined in age;
open to listen to the sighs of pain,
and tears of grieve;
open to give money to fill a need;
and open to give love so others might hope.
She's always so kind and thoughtful
so obliging and tolerant,
so sensitive and sympathetic,
so slow to anger and quick to forgive,
and so generous to those in need.
She's more than willing to give of herself
including accepting and loving others

and supporting and caring
for those who may be different,
and all with open arms
and an open, willing and loving heart.
And she volunteers most of her time for her church
driven by a single purpose that she might,
in her small way,
leave behind a better world for our young.

And on this your very special day,
my dear friend,
I like to send you
my warmest birthday greetings
and tell you that
your compassion and kindness towards others
is an inspiration to us all.
Thank you for your many years of friendship,
for who you are
and all that you do.

March 2021

For Kathy
A Giant of a Zontian

When I first joined the Club
some six years ago,
I was petrified to learn of my first task:
to write up a pamphlet introducing the Club,
in both English and Chinese
when I knew practically nothing about it.
The President comforted me:
"Ask Kathy.
She knows everything there's to know."
And indeed she told me everything
I needed to know,
and this
"Ask Kathy"
has been the refrain of the Club
for all these years since I know her –
for every situation
or state of affairs,
be it big or small,
trivial or significant,
incidental or crucial,
a gossip or a joke –
we ask
and we get the answer,
promptly.
She's the encyclopaedia of the Club.
Her knowledge is across the board,
in-depth and meticulous,
comprehensive and complete
and cross-border,
from the rules and bylaws of every Club
in addition to those of our very own,
for there are 7 in Hong Kong,
and those in our region;
the history of the organization,
the various advocacy and service projects,

the regional committees and sub-committees,
to the achievements and awards over the five decades,
complete with the dates
down to the exact months and days.
She's our walking Kathipaedia.

She shares her information and data
willingly and readily,
for her generosity of spirit is unparalleled.
She wants to make sure
that members have the correct information
to make the right decisions
instead of wasting time
and wondering what should be done
when they could start right away,
taking the opportunities
to collaborate
and get things done quicker and better.
She has all the know-how
and wouldn't let all that accumulated experiences
go to waste.
She takes the time and effort
to share them with us,
whether highs or lows
and takes on a mentoring role
to help others achieve
the success that she has already accomplished.
She is generous
not only in giving information,
but credit, time, expertise,
and the benefit of the doubt,
openly and frankly.
She's outspoken without beating about the bush
or mincing her words.
She might sound fierce

particularly to the new comers,
but she's a paper tiger.
She appears strong and powerful
and even threatening,
but she has the tenderest of heart.
I have witnessed it many times
when she would quietly and successfully
resolve conflicts among members;
and I too have experienced her kindness
on two particularly delicate situations
and for that I am most grateful.

But above everything else,
Kathy is a leader.
"Leadership is influence – nothing more, nothing less."
She is able to translate vision into reality.
I was invited to chair the Advocacy Committee
a few years ago
and proposed to tabulate
a Hong Kong Low-income Working Women Wellness Index.
She listened actively with great attention and sincerity,
and was not only receptive and open to new ideas,
she's innovative and thought-provoking
and motivated us to think out of the box.
She believed in us
and persuaded the Board
to give us a resounding thumbs up endorsement.
She inspires us
with her principles without compromising;
she refrains from making false promises or taking shortcuts,
and chooses options over personal gains.
Everything she does is for the Club
and never about herself.
As a Parliamentarian,
her primary goal is to help all the Clubs and their members

succeed in their missions:
from promoting equal access to education,
health care and equal economic opportunity,
to safeguarding legal rights
and freedom from violence,
she wants to make a difference
in the lives of women and girls everywhere.
She always keeps a positive attitude
and cheerleads many of our initiatives;
but while she keeps a tab on all,
she stays in the background and let others shine.
She is inclusive of participants
and welcomes new members whenever possible.
She wants everyone
to feel that she's a member of this big family
and part of every of its commitments,
mission and overall success.

What's perhaps most inspiring about her
is her dedication to the Club.
She is known for getting things done,
and on time,
and is greatly responsible
for developing and maintaining our Club's position
as the first and premier Zonta Club of Hong Kong.
Apart from all the important awards,
too numerous to count,
that she has received throughout the decades,
her yearly 100% Attendance Award
shows more of her commitment and devotion
to the Club than anything else.
One sees her in action everywhere:
she's here and she's there
and she's everywhere;
and she's punctual at all times,

for luncheons,
meetings,
fellowships,
conventions,
forums
and all other activities.
Her visible passion for the Club is infectious
and inspires others,
particularly our younger members
to follow what she believes in:
"Never doubt that a small group
of thoughtful and committed people
can change the world".
And this is what we all believe in
and strive to achieve.
Let's all,
under your leadership,
Kathy,
serve with love,
compassion,
gratefulness,
humility,
and with joy in our heart.

March 2021

For Charmaine
A Woman with a Heart of Gold

There's this young woman I know,
she's so bubbly and full of life,
so enthusiastic and warm
she could melt even the iciest of hearts.
She makes a conscious choice
to set a baseline standard
for what she will accept in life,
as she knows how easily
one could slip into habits or attitudes
that might yield a quality of life
that's dull and commonplace.
She's never the one
who would have her lunch
on a tupperware
even when she were to have it on her own.
She will set her table
as if she's having guests
not forgetting
whatever little floral arrangement
she could manage
to be a part of it.
She creates her moments,
for she knows that life
is made up of small moments like these
for most of us.
It takes wisdom to appreciate
that joy comes to us
in ordinary moments
and she is not willing
to risk missing out
while striving and struggling hard
chasing down the extraordinary,
which might never come.
Yes, the quality of life is more important than life itself,
and that's how this young woman lives her life

and why she's such a blessing to those around her.

And she's the most resourceful woman I know.
You order a 6-feet Christmas tree
and it's a feet too short when you put it up...
No worries –
she will get you a box of a stand
surrounded by almost the same leaves of the tree
and your now 7-feet tree looks nothing amiss
even before you put the Christmas gifts around it.
Your leather skirt is stained
and she will get you a matching embroidered pin
to cover the mark.
Video not functioning? Hand cut? Arm fractured?
Here's the expert to help you so she tells you.
Want to tailor-make your own stationery?
4-5 simple or elaborate designs on your screen in an hour.
You choose the best Christmas or birthday card
for your dear friend
and it gets smeared or split
while you are writing your special line?
Send her a photo and she will get you
a complementary floral or animal
or graphic design self-adhesive ribbon
to make your card look even more elegant.
She has everything you need
to sew or glue,
draw or colour,
bake or roast,
print or fold,
toss or fly,
slice or tape,
whatever you fancy...
just tell her and she will give you the tools and the expertise
from her good-self or the right experts to provide them.

And I have long realized
that it's not resources but resourcefulness
that ultimately makes the difference.

She has such a generosity of spirit too.
The Buddha said:
No true spiritual life is possible
without a generous heart.
Generosity allies itself
with an inner feeling of abundance,
the feeling that we have enough to share.
Yes, she has learned,
just like another friend of mine,
that happiness is not made
by what she owns
but what she shares;
nor is it by how much she gives
but how much
love and joy she puts into giving.
I know for I have been receiving
only once too often...from you.
I would never forget
the sharks fin you took days to prepare for me,
the fresh leaves to decorate my plates
the delicious beef slices to be my appetizer,
the Perrier Jouet Champagne glasses to grace my table...
but above everything else,
your love and joy
when these were sent to me.
Your joy is my greatest reward,
so you would say when I thank you every time.
You have mastered the art of giving
and this positive energy is contagious.
I hope there are more others like you
who could inspire others

to go out and plant seeds of happiness
through giving,
and that you will be their guiding light
to truly experience how sharing
is what makes life meaningful.

February 2021

For Susie
A Woman Who Stays True to Herself at All Times

Here's a professional woman
who is gracious by temperament,
elegant by disposition,
fashionable by intuition,
and genuine by nature.
We ran into each other twice,
once while shopping at the Landmark
and another at the HSBC Premier Lounge,
only knowing of each other
through distant association,
but kept our words a few weeks later,
as we parted with our little
"must get together sometime"
when such banality is rarely taken seriously
by most,
including me and her,
and began a friendship
over the years
that we both wish
to nourish and treasure.

The very essence of her being,
stood out very much in evidence
only after I talked to her
for less than an hour.
She's one of the most genuine persons
I have met and liked,
and proved it
through the years
as we gradually become good friends.
She's a perfectionist
in all things
that matter to her
and couldn't care less
if they don't.

She's peculiar
about how her meals
kitchen-wares,
skirts, dresses and pants,
books and magazines,
sofas and cushions,
Lalique vases,
Baccarat crystal collection,
and every home item really,
that should be prepared,
placed or done,
and would be dismayed
when they are not.
She's very selective with friends,
and while generously spending time
with those she likes,
would waste not a minute
over anyone
whom she doesn't.
She's least affected
by others' reactions
which she believes
reflect more of the character of others
rather than hers.
And just like what she would always say:
"don't confuse my personality
with my attitude,
my personality is who I am,
and my attitude
depends on who you are",
she has helped me
set boundaries
for unwanted relationships
that I have miserably failed
only once too often,

and for that I am most grateful.

She is indeed most genuine
with every quality that goes with it.
She is honest, truthful and sincere
in the way she treats friends.
Her emotions are real and heartfelt.
She's true to herself
in the way she thinks and feels.
She speaks her mind openly,
and is comfortable
presenting her ideas
without expecting or needing
to convince others that she is right.
We are able to happily spending time,
sharing and exploring
thoughts and ideas,
goals and targets,
beliefs and ideals,
standards and aspirations,
serious or trivial,
sublime or commonplace,
grandiose or otherwise,
as we know
we are going to get response
truly from the heart
which could very often
give us guidelines and perspectives
for meaningful communication and self-reflection.
Her readiness to admit her faults,
oversights or mistakes,
enables her to be less judgmental
and more receptive of those of others'.
Her fundamental assumption
about human complexity

and her reluctance
to view people through the lens
of preconceived expectations
allows her a purer frame of mind,
that usually leads to
more direct and honest
interaction with each other.
It's always a joy to attend with her
an art exhibition or a fashion show,
a seminar on South Sea pearls or antique jade,
or lunch together,
be it simple or elaborate,
when we could talk about anything under the sun,
comfortable and at ease with each other.
It's like an autumn breeze,
a walk in the park,
a bed of roses.

She has over the years
developed a sartorial personal style,
that is as unique as she is.
She clearly knows what suits her
and what style feels most true to her.
Style is as much about knowing yourself
as well as knowing
what you want to portray,
so she told me.
She first started
with a basic understanding
of what kind of body shape she has
and what works for her.
It's only when one understands
one's strengths and weaknesses
that one could buy the right clothes
to either accentuate

or camouflage them.
Hers is one of simple minimalist
but classic and sophisticated.
She goes for neutral colours
like white, beige or cream,
navy blue, black or grey.
Sometimes she goes monochromatic,
featuring different shades or tints
which allows her to vary the same colour
but remain classy.
And how she accessorises what screams personality.
She usually wears one single statement piece
and her favourite
is large well crafted
and brilliant and translucent jade pendant
strung up by jade beads
or silk chords
or refined lace
or intertwined braids
that always takes my breath away.
There are other little accessory tricks
that she brings into play
to charm and delight me
like throwing a fetching scarf
over a cashmere sweater,
wearing a statement brooch
or a rare purple jade bangle
or simply carrying a timeless limited edition handbag.
She never goes overboard with accessories
and to her
they are the cherries on the cake,
capable of transforming basics
into a fashionable ensemble.

It must be almost

ten years since we first sat down
for lunch or tea
and it seemed most unlikely
that we could become friends,
and such good friends too,
when we appear so different in every way,
only to find
that we do share similar core values in life,
and many interests and perspectives to boot.
Friendships develop in a million different ways
and all good friends strive to achieve the same goal:
to be a source of love and support,
which I have since received from you.
Thank you for the unique bond
we have for each other
and this little tribute
is my way to tell you
how much I enjoy and treasure
your friendship.

February 2021

She's born into an eminent family
and with a great-grandfather
featured in our history books,
it should easily explain
how our lady
was raised up
with the art of being a lady
that goes far beyond social etiquette,
table manners,
fashion relevance
or how to set the dining table properly
for different occasions,
to become what *she is* now –
the ultimate lady of class,
in the way she is
and carries herself.
And so the saying goes,
"Being female is a matter of birth,
being a woman a matter of age,
and being a lady a matter of choice"
and in this case,
it's an unintentional choice
for our gentle lady.

What she impresses me most,
over the decades
since I knew her,
is her natural composure.
She has such a beauty of self-control
which allows her to stand tall with grace,
when facing conflicts,
challenges,
and even loss;
and not to be boastful
when it comes to successes.

She has taught me
that the less one reacts,
defends,
explains,
becomes fearful or worrying,
the more command one has over a situation.
It allows one
to maintain a certain stillness
in the face of life's constant changes,
and to become more receptive
to the invincible truth
that *less is more*.

Then there's the kindness of her spirit
that's as enlightening as it's uplifting.
Kindness to her
is a way of life
which transcends
being merely amiable and polite,
dependable and reliable,
considerate and non-judgmental,
easy going and sympathetic...
it is being able
to offer help to others
who are needy
without giving a second thought
and without thinking of herself.
It is when she gracefully
allows others to shine
and have their moment.
It is when she doesn't contribute to dissension,
discord or division among friends,
and stops it
when others stir up drama
in her kind and gentle ways.
My lady is a vessel of mediation and peace.

She has a very rich and fulfilling retirement life
through volunteer work for the community
which is like a fulltime job to me,
setting an exemplary benchmark
for all of us to follow.
Volunteering allows us
to connect to our community
and make it a better place for all of us.
Even helping out
with the smallest tasks
can make *a real difference*
to the lives of people and organizations in need.
It is one of the best ways
to help benefit the public
or give back to our community.
Our lady works
for various museums and schools,
advises and sponsors scholarships for gifted students
and sits on school boards
to make important decisions
that would improve
the quality of school lives of students
and their future careers.

But perhaps what's most inspiring for us
is her artistic talents
unrevealed to us
throughout the years
which was but discovered accidentally
and only recently.
We all know how well she draws
which she has done on stones,
wine bottles and greetings cards,
glasses and boards,
and practically every possible surface,
she could find,

be it rugged, irregular, bumpy, rocky, thin, sturdy or smooth;
but to have such a gem of a talent
among us
without fully realizing it
is shameful if not criminal for us.
I was sitting next to her
during a concert
when she took out
her notebook and drawing pen
and started listening to the music with her ears
and sketching with her heart,
(just like our other friend who multi-tasks at all times)
I noticed it soon enough
but couldn't figure out what she's doing
as her eyes
danced from her piece of paper to the stage
and back again,
again and again.
For over an hour,
as I kept sneaking glances at her sketch,
all I could see were
numerous dark black oval shapes
on top of elongated tubes
with smudges and dots,
stains and smears,
blotches and stains
here and there
and everywhere...
I thought that's not really the concert for her
and she's drawing for her granddaughter.
Then as she started to apply,
after an hour or so,
a full range of tones
from black,
through a multitude of greys

to off-white and white,
I was enthralled
all of a sudden,
as my eyes feasted on a rich visual image
of the orchestra in the front,
the choir at the back,
and the musicians with their instruments
in the centre,
complete with a variety of hairdos,
facial expressions
and even glass rims –
the painting of a full orchestra and choir
with the conductor very much in action,
waving his baton.
It's hyper-realistic
but it's the beauty
in the smudges
and irregularities
and artistic interpretation
where the soul sneaks in,
and we were spellbound,
inspired and in awe.
Our talented lady
is neither shy nor self-conscious,
she just likes to keep her light
under the bushes,
and to our great admiration,
so free from vanity.
She not only paints beautifully,
her calligraphy
in both English and Chinese
is phenomenal.
There's such a fierce persistence in her art,
so adventurous and versatile with her pen,
so intuitive and fearless in her reflection

and yet so quiet and gentle in her demeanour,
one cannot but like and admire her,
this our modern lady
of class and grace.

January 2021

For Sharon
A Woman Who Lives for the NOW

Life is too fleeting,
too incidental,
too miniscule,
so she would tell me,
only once too often,
particularly at times
when I am complaining over a grievance,
expounding a specific theory,
protesting against a decision,
arguing over a difference of opinion,
responding to an unfair criticism,
explaining a contention,
or lamenting something lost...
that you can change your life or not change it,
it really doesn't matter at all:
for in the end,
when God calls,
we simply need to go,
whether it's too soon or too sudden,
too unexpected or too unready,
but really,
whether we like it or not.
"Live for the Now"
is her life motto
she religiously lives by every single day.
"No regrets"
and that's how she believes everyone should live,
"Don't let life pass you by
waiting for things to happen",
said my philosopher,
"create your moment
and live every moment to the fullest".
Let all petty arguments go,
the distress of a job loss,
the wretchedness of a broken heart,

or even the pain and grief
of the loss of a loved one...
The agony feels like it will last forever,
yet it will inevitably leave
just like everything else –
so stay patient and wait,
for time will always heal.
Life has no meaning
until you make it so.
Waste no time before it slips away,
for life is but a fleeting moment
and done and gone
in the blink of an eye.
And so let's learn how to dance in the rain,
my friend and thinker said,
grow stronger
despite the disappointments and challenges,
learn to enjoy all the little joys that life will always bring
if we open our hearts,
and play and laugh and have fun
whenever and wherever we can.

You will be most mistaken
if you think she's frivolous,
heartless or even superficial
for she's one of the kindest,
caring,
compassionate,
considerate,
generous,
knowledgeable,
helpful and supportive women I know.
I know for I have experienced it myself
again, again and again.
She works hard and is still working as a family lawyer at 70,

just as she plays hard
and enjoys everything she does every day –
she loves playing card games
and sport activities of every possible kind.
She's an avid golfer
and plays even at the height of the pandemic
with robust anti-infection devices
and belongs to I don't know
how many golf clubs or golf groups.
To her golf is the closest game to the game we call life.
You get bad breaks from good shots
and good breaks from bad shots,
but you have to play the ball where it lies.
You will be awed to see
her colourful bling bling masks
she made for golfing
and the costumes for her team members
she prepares for the merrymaking
with every winning celebration.
And she's not being lovingly nicknamed
Lady Bling Bling for nothing –
feasting us with such sensory pleasures
every time we see her,
with her one-of-a-kind hot-pressed rhinestone patterns
on her dress or pants,
T-shirt or jumper,
scarf or neckerchief,
shoes or sneakers,
purse or handbag,
hat or hairband,
or anything with a surface.
She loves dancing,
and has choreographed several dancing routines
for our Zonta Clubs
to mark several of our special occasions,

and now everyone,
both the dancers and the audience,
simply love her –
for her artistry and creativity,
physical stamina and athleticism,
discipline and persistence,
and above all,
interpersonal and leadership skills.

So how could two
so apparently different persons
become such good friends?
She doesn't take every word to heart
while I am very sensitive.
She's the present,
in the now,
while I worry
and wait for the future.
She loves sports
while I am a city girl avoiding the sun.
She's naughty and fun-loving
while I am serious and traditional,
She loves dancing
while my legs can't tell left from right...
yet as we get to know each other,
we find we do have much more in common
than we would never have imagined
should we stay away from the start.
We are both intellectually inquisitive and literacy curious.
We are avid readers,
and share many of our favourite authors;
we could talk about
a piece of literature or a serious subject
for hours on end
and it's still never enough;

we enjoy the same Korean or Chinese drama series
and have had such fun
reliving even the trivial tiniest details
after we finish watching them;
we go to see movies together,
and I have for the first time in my life,
seen each and every one of them with her
without knowing what I would be seeing,
as her naughty disposition
wants to surprise me
and "not to spoil my fun" as she so claimed.
But what really brings us together
is our shared spiritual beliefs
and similar family core values and attitudes to life:
respectful to authority,
caring and generous to the underprivileged,
compassionate and kind to the needy,
empathetic and obliging to those in pain
and helpful and kind to friends,
particularly during bad times like now
as we have never experienced before,
when the whole world is suffering,
when the both of us
have the same urge
to offer ourselves
and do whatever little we can
to improve the world we live in
by getting actively involved
in community service work
or whatever it takes us.

Thank you for your friendship
Lady B,
which I hugely cherish.

January 2021

For Barbara
Crème de la Crème

No, no poet with any sense
nor poetic license
would be induced to write
about a lady of her stature.
No, one should have the intuition not to do it
even when brutally coerced
or handsomely paid,
as how could one begin to write
about a prima donna
of such eminence
who is at once an investment banker,
regulator,
business leader,
philanthropist,
entrepreneur,
senior advisor to international boards and commissions,
published author,
economic and political analyst and critic,
linguist,
conversationalist,
humorist
and philosopher
with a life style that advocates health
and teaches by example
how one could lead
a bountiful and meaningful life
and live to 100?
No, that would be too daunting a task
unless one finds one or two aspects in one's life
when one might share the same wavelength with her,
no matter how incomparable.
No, I could find none
no matter how I tried,
and in desperation
to be of service to my dear friend,

I jumped in,
myself forgotten,
my mind all set,
and my heart open
to her universe,
and if you are ready,
I am:
here's the tale
of an exceptional lady...
Our lady is one of the most knowledgeable women I know:
she's not just well educated
which proves to be nothing
for not just a few,
she's intelligent and perceptive,
sharp and intuitive,
innovative and entrepreneurial,
far-sighted and shrewd,
always eager to learn more,
always interested to experience
and try new things,
always keen to stay updated
on new developments,
always ready to offer valuable advice and suggestions,
for like Tennyson's Ulysses,
she's a seeker,
a seeker for adventure,
experience and meaning
to render her life even more meaningful.
Ulysses faced several challenges
in his life,
and now that he's back in his homeland
living a peaceful life,
the uneventful life of a monarch
who only settles disputes
does not appeal to him.

Our lady too would never settle for less,
she too aims "to strive, to seek, to find, and not to yield";
and now at 78,
she's still very much involved
across a broad range of sectors.
She just told me
about QAnon zealots
who believe Donald Trump
is secretly still President
and carrying out executions at the White House,
when I,
considered well informed
by any standard,
have never even heard of such wild new theories.
She's the quintessential walking encyclopedia
interested in politics, history, economics, finance,
and basically everything else.
She's an inspiration.

She might be a serious philosopher,
yet she could be so amusing and entertaining,
quick-witted and humorous,
when it comes to storytelling time –
for our heroine is the queen of words,
and she lives and breathes amid a wealth of them.
And that's why I so much look forward
to joining my husband's lunch or dinner with her,
patiently waiting for agenda AOB,
to be amused,
entertained,
shocked,
bewitched,
by anecdotes big or small,
trivial or profound,
ugly or beautiful,

honourable or disgraceful...
listening to all these
while feasting my eyes
on her animated face and her expressive eyes...
and the warmth in my heart
would linger on,
as I wake up the next morning
remembering her tale about a kleptomaniac
from a family of old money
who got caught
not while he was stealing a VCD copy
but when he was returning to exchange it for another
as he did not like the one stolen;
or her serious advice
that one could disregard all the Ten Commandments
except the 11th
which is not to get caught;
or her picturesque description
of a fallen mandarin's taped conversation with God
or her entertaining impersonation
of an Indian billionaire
sitting between her and another Hong Kong billionaire,
nonchalantly asking him:
"I am a billionaire,
you are a billionaire,
is there anything we could do together?"
Her life motto:
work hard and play hard.

Yet what really tugs at my heartstrings
is when she very graciously
shared with us
those magic moments
she has herself experienced
like meditating

while sitting alone on a rock
facing the sea in Brittany
when the whole world was all hers
or watching a spectacular sunset
in silence for a few minutes
alongside with one of Hong Kong's best and brightest,
but busiest,
just the two of them.
Yes, she might be overwhelming at times,
and could be merciless,
she atones for her occasional overkill
by being always
so refreshingly delightful,
so utterly convincing,
and so discerningly quick-witted,
one is touched
by her fiery command of the language;
her enormous wealth of information and knowledge
on everything and anything
under the sun,
the moon and the stars...
her keen interest in physical and mental health;
her ardent enthusiasm to live life;
and her conscientious dedication
to fill her life journey
with such joy,
excitement,
satisfaction,
purpose
and good old fun.
She's a gem to all her friends
and anyone who chances
to attend her seminars
or listen to her talks and lectures.
Thank you Barbara our dear friend,

For Barbara Crème de la Crème

for the joy you bring us
and your precious friendship
we both treasure and cherish.

December 2020

For Chris
A Woman of Courage

"It's impossible", said Prudence,
"it's risky", said Experience,
"it's pointless", said Reason...
"Give it a try",
her heart whispered,
and she did,
again and again,
she did,
over the years,
well into the stillness of the night
when the mind is clear
and the heart heavy
with a mission impossible
to be accomplished.
Resolute and firm,
she rushes in
where the weak and the selfish
fear to tread,
churning out controversial articles
two or even three a day,
to deafen
the wicked and the immoral,
while getting
bullied and threatened,
exposed and even terrorized...
Yet she continues,
unafraid,
for she knows that
she's not given a magic pen
for nothing:
she has to put it to good use
and she has.

She is YOU –
A warrior,

a fighter
and a champion...
We all see it
and we are grateful.
Heavy is your crown
and yet you wear it
not on your head
but in your heart,
and humbles us all
who stay
and will stay
behind you forever,
with the loudest cheer
in silent joy and gratitude.

December 2020

A Birthday Tribute to Ha Ying
A Woman of Substance

She said
the other day,
"I first saw you forty-one years ago..."
and I swallowed the temptation
to ask her for the exact date
for I know she would have remembered.
Here's my dear friend,
the fictional Funes the Memorious,
in real person,
who can remember everything –
be it big or small,
major or minor,
important or trivial,
that has happened to her
and the people around her
over the course of her life and others.
It may be in her genes,
or developed
from her massive grey matters
this magnificent memory of hers,
yet it is but only one of the
many gifts that God,
to my huge envy,
has bestowed on her.

For she's a tower of intellect,
possessing a great capacity
for thought and knowledge,
not only in her own medical field,
but in so many other areas,
with an intuitive understanding
of a wide range of subjects –
that of art, history, music, dancing,
politics, philosophy and humanities,
along with a razor-sharp sense of perception,

defying traditional notions
that a devoted wife and mother
could not equally excel
as a brilliant and scholarly mind.
She might appear quiet and reserved,
but she is fire,
a restless, free ranging eclectic
always at her 'amateurish best'.
A woman of substance,
she's gifted with a lively mind,
rendering every of our sharing so inspiring
that I await in anticipation
every time
for our next.

A woman of class,
she has a distinct sense of style,
displaying self-respect and appreciation,
dignity and grace,
moderation and kindness
in her daily interaction.
She keeps her promises and obligations,
and is known among us
as someone who is dependable.
She's supportive
to friends who are in need,
and worthy social causes
which aim to serve the community,
and will offer help
in whatever way she can.
She's never the one to seek attention,
and avoids it even when it comes glaring at her face.
She could be critical at times,
insistent of what is right from wrong,
but she's never mean,

in both words and deeds.
Here's a girl with a mind,
a woman with attitude
and a lady with class.

Her life motto is carpe diem:
seize the day,
and enjoy the present.
She has taught me
by examples,
how not to worry about the future,
for a minute worried
is a minute wasted and gone.
She creates her moments
for she knows that life
is made up of small moments,
and big ones occasionally,
and these personal choices
she makes every day
lay the foundation
for her mental and physical health,
but most of all,
determine the quality of life
she leads for her and her loved ones.
To her, it's not the years in her life that count,
it's the life in her years.
Yes, the quality of life is more important than life itself,
and that's how she lives her life,
and why she's such a blessing to those around her.

She's also most generous.
She has learned
that happiness is not made by what she owns
but what she shares,
nor is it by how much she gives

but how much love and joy she puts into giving.
I know for I have been receiving
only once too often.
I might forget as time goes by,
the many boxes of Lane's exquisite shawls,
the D & G 3-D rose brooch,
the Shiatzy Chen padded embroidered coat...
but I will never forget
your love and joy
when these were sent to me,
making me at times
envious of the gift you are blessed with
that few would never have the chance to experience.
So let me take this opportunity
to express just how much
I appreciate your kind generosity
to me and my family,
and that it may return to you a hundredfold
not that you are asking for it.

And so on this your very special day,
may you continue to bless others
with your exceptional gifts –
to make a difference to this world
that's so fragmented,
so filled with hatred and anguish,
that we must work together
and give whatever little we can
to have it
healed and rebuilt.

Happy Birthday,
Ha Ying

Best,
Julia
November 2020

A Birthday Tribute to Maria
A Woman with a Personal Calling

She's a leader,
like she's born to be one.
She has a personal calling
that's as unique as a fingerprint:
devoting her whole life to public service,
and expecting nothing in return.
In her mid-30s,
she became a member
of four different levels of representative councils
in Hong Kong
when she was appointed
to the Executive Council,
the top advisory body in the government,
unprecedented and unrivalled,
a political legend
that's remembered and much talked about.

She's a warrior,
strong and decisive,
she never takes no for an answer;
independent and perceptive,
she always gets the job done;
forceful and determined,
she stands up for what she believes in.
She confronts every challenge with dignity,
weathers every storm with conviction,
acknowledges every defeat with humility,
and won't let anyone nor anything stand in her way.
Today she could look even a threat
dead in the eye
and give it a wink,
and is now the Vice-Chairman of the Basic Law Committee,
the highest advisory body on Hong Kong's mini-constitution.

But above all else,

she's a friend,
a true friend one can only dream to have...
generous and supportive,
she's blessed with the gift of giving
and a willingness to help others.
She reads character,
and recognizes competence and potential,
and has the bigness of heart
to genuinely say:
"I admire your qualities and attributes",
inspiring us to be the best we can be
and providing opportunities for us
to shine
and to serve,
through her connections and network.
I am blessed to be among those
whom she trusts,
and have lost count
of the number of people and projects
I have known and participated
through her.
She's always so modest and kind,
and helps the less fortunate compassionately.
I know
for I was there,
not just once,
but many times,
a witness of her generosity
in every sense of the word,
offering most often her precious time
which she has so little.
She has touched the lives of many –
from the needy, the deprived
to the confused and agitated;
from the destitute, the persecuted

to the exploited and oppressed;
and the Police in particular,
the most abused
since last year's social unrest,
with her light shining through,
brighter than the morning sun
and the nightly stars,
giving hope
that there might be a better tomorrow
as she travels back and forth
from one meeting to another,
from the dark alleys to the community centres,
from Hong Kong to Beijing
whenever and wherever she's needed.
So now you all know who this wonder is
much as she may seem too good to be true:
Yes Maria, this amazing woman is YOU!
For you are more respected than you think,
more admired than you reckon,
and more esteemed than you will ever imagine...
and on this your special day,
let me send you this little tribute
so you know how inspiring you are
and how blessed I am
to have a friend as fine as you.

Happy Birthday Maria.
Have a blessed day
filled with joy and love,
and everything else
you so deserve.

16 October 2020

A Birthday Tribute to Amarantha
A Woman with a Difference

Like to meet someone special,
an old friend asked,
why not,
I replied.
It's as if fate couldn't wait,
and tea was arranged for the next day.
My friend left soon after
but we stayed on –
on and on
we stayed,
and what sharing we have had,
covering almost everything
under the sun
from political, economic to social issues,
from personal beliefs to secret aspirations,
for hours on end,
like two old friends
who have known each other for life.
For it's not like
one could easily find a kindred spirit,
and when one does,
one does not let go...
And eight years on,
this connection
which first put us together
not only remains,
but grows stronger,
deeper,
and more fulfilling,
by the day.

A woman of intuition,
she's gifted with a lively mind
for a keen and quick insight
and a direct perception of truth and fact.

It's such a joy to be able
to communicate with each other
without having
to state the obvious
and be understood
even when unexpressed or barely implied.
For she's not a family mediator
for nothing –
she's an effective communicator
and an active listener too.
She reframes her words
before they are out
so whatever is expressed
is light and grace,
decorum and balance,
stimulus and sensibility,
and good old common sense:
for she's neither ice nor fire,
but everything in between,
rendering every sharing
such a comfort and fellowship,
a solace and delight,
and an awareness and enlightenment,
and I await in anticipation,
every time
for the next.

A woman of strength,
she's gifted with an innate thirst
for knowledge
which is both empowering and nurturing
to those around her.
After taking up a leadership role,
she soon realizes
that the new age

is demanding innovation
at such a rapid rate
that endless ideas are needed
in order to have a job well done.
So instead of telling everyone
what to do,
she motivates others
to come up
with the best and brightest ideas
that have never been thought of before.
She holds back
from giving her teams
the answers
but states the problems
and let them explore and create.
She inspires by example
and lives the person
she wants others to be –
one of hard work,
discipline,
determination
and commitment.

A woman of faith,
she's gifted with an incredible capacity
to endure pain and suffering;
to stay positive in sickness;
to embrace challenges in times of difficulties;
and grow strong by reflection.
Every adversity brings her new experiences,
and new lessons
which she fearlessly faces.
For she knows that God has a Grand Plan
and with God,
we are stronger than our struggles

and more fierce than our fears.
And that God provides comfort and strength
to those who trust in Him.
She fights her battles through prayers
and wins them through faith.
That's how she manages
to smile and even laugh
with every challenge in life,
and live fully in victory,
showing us that
"if one has faith, one has everything".
And with faith,
comes love and compassion,
kindness and charity,
virtues and humanity
as she opens her arms to the poor
and extends her hands to the needy.
touching the lives of so many.

You may wonder who this woman is
for she may seem too good
to be true:
so let me shout aloud
in case you still don't know –
this amazing wonder
is You!
Yes, she is You:
For you are braver than you believe,
stronger than you seem,
smarter than you think,
kinder than you know,
more generous than you reckon,
and thrice as beautiful as you have ever imagined...
and on this your very special day,
let me send you this little tribute

so you know how inspiring you are
and how blessed I am
to have a friend as fine as you.

Happy Birthday
Amarantha,
We love you!

May 2020

For Stephen at 70
Our Very Dear Friend

And so,
Adam and Eve
live happily ever after,
in the Garden of Eden
where flowers bloom in all seasons,
where trees bear fruits
of every possible kind...
with oranges,
without one single doubt,
the sweetest
and the fairest of all;
the absolute favourite
of our Prince Charming and his Love,
that they are
so much nourished
by its simple pure goodness,
that they are empowered
with a strength
so dynamic,
so vitalizing,
so electric,
almost mystical,
that our Lady
continues to dazzle us,
with her kindness and humility,
despite being
the ultimate superwoman
up in the corporate stratosphere,
multi-tasking
with the aid
of her increasingly more
surgically attached devices,
while helping others in need
with her infinite resourcefulness;
and her Prince Charming

never ceases to delight,
nay,
charm,
fascinate,
bewitch everyone
who knows him well,
with his inborn childlike sense
of wonder and possibility,
shining forth
ever so gently and quietly
from his beautiful innocent heart.

Our understatement,
and so that's how he is known to everyone,
rarely speaks,
he just sits there,
with his lips curiously curled up
in his signature smile,
one hand holding his wine glass
all ready for a sip
and another
conscientiously pouring wine
for every half-filled glass,
making sure
all his friends will not go in want
of such sweetness from heaven...
He listens
actively and with patience;
and when he speaks,
which he rarely does,
the world stops:
all heads bent forward
with ears open
for the words
that would come slowly and gently:

he once told us
how he was
standing inside an MTR train
and patted shyly by a little girl
with her finger pointed to an empty seat
vacated by her mother
for him.
"This is the ultimate parental teaching
by being a good role model
for our children",
he said,
"and my heart trembles with joy
to witness such beauty".
He also told us
how his heart went soft with emotion
after a concert where he's a part
when the mother of a handicapped young girl
came up to thank him
as her little girl,
not easily happy,
was happy
listening to the music.
"To know that a little girl is happy,
even one little girl,
makes every of our efforts worthwhile"
he affirmed.
We certainly do not forget
his infectious excitement
when he saw the bamboo decanter
in a Japanese restaurant
for the first time,
which he has remarked,
in awe and with admiration,
again and again
more than ten times

over dinner.
"This is the first time
I have ever come across
a bamboo decanter
when I have been drinking
for well over five decades!
This is art in its purest form"
The beauty he sees in simple everyday objects
and the joy he has experienced
and shared with friends
is as moving as it is inspirational,
and touches our soul
which without our knowledge,
has become conditioned, insensitive
and simply too tired to feel.

And so on this your very special day,
let our soul
clap its hands together
and thank you
for what you have
unknowingly
shared with and taught us,
all these years.
Let's all look for the inner child in us,
wake her up from her long scared slumber,
embrace her,
and tell her that it's safe
to come out and play now
"for we shall carry you
in our loving arms,
and no one
can and will
hurt you
any more"!

It has been such a privilege
to be with you
for the past decade,
and we look forward
to our next.

Happy 70th birthday,
Stephen,
WE LOVE YOU

January 2018

For Edith
Congratulations for an Appointment So Well Deserved

No, you won't let me leave tonight,
If I don't bring to light,
This success story in black and white.
So my dear Oranges, do sit tight –
Here's the narrative I shall now recite,
Starting with a tale of three cities if I might...

In late 1999 Orange came into sight,
Of Mannesmann, the world it did excite,
To buy it for a hefty £18 billion, yes that's right,
And when Superman gave the green light,
Our comrade had to take an urgent flight,
To Dusseldorf she went, armed for the plight,
Her heart was heavy, her luggage light,
All possible contracts she held tight,
Plan A-Z, all ready for the fight –
It took many a day and many a night,
Alas the bark was loud, but not the bite,
The deal was signed and sealed alright...
And yes you are right, behind this Orange rite,
Was our very own Orange so brainy and bright!

And so she climbed to further height,
One achievement after another she stole the limelight,
From one jurisdiction to another the map she did rewrite,
She could fly for 24 hours and remain a dynamite,
She has resilience, she has foresight,
She's Superman's secret weapon and open delight!
With this new appointment coming to light,
We are so happy for you but just not quite,
With more transactions you now need to expedite,
Every process, every task, every fight,
You used to toil from sunlight well past midnight,
Now your time table you need to rewrite,
with every schedule so tight!

No, your surgically attached BB could never be out of sight,
Our doctors will help insert more, to cope with any sudden blight.

Congratulations, our very own Orange bright,
You are a winner but your health please never, never slight.

December 2016

A Birthday Tribute to Jerry, Our Dear Brother in Christ

When I first knew him,
some six or seven years ago,
he's a changed man,
and so I was told.
It started with his fear of God,
this our friend,
and the fear of God,
and the fear of God alone,
has removed his self
from his heart
replaced sacredly
with Jesus on the throne.

Yet he never wears his faith
on his sleeve
and one can hardly smell it
even when one sits right next to him.
He stays the man he is
to the outside world
and his heart
he has it hidden
from most others
except those
who have received and received aplenty
from him.

What a jolly man he is,
always so playful and in high spirits,
so mischievous and so full of fun.
He is forever so bubbly
and you will have to kill him
to stop him from opening his mouth;
and how he opens his mouth
wherever and whenever he can.
A joke here and a joke there;

a trick here and a trick there;
a funny story here and a funny story there –
all with comical stunts in between.

How he would tease you
whenever he has a chance.
There is simply no escape: he notices everything:
from the ridiculous to the sublime,
from the mundane to out of the world,
and from the trivial to larger-than-life –
there is always something
bordering the hilarious for him to say.
He has a vocabulary quite his own,
with words coined by himself and understood by none,
and when you are his unfortunate target,
he makes you laugh and cry all at the same time.

How he teased me on that miserable afternoon
when I most broken-heartedly lost
my unique Moschino red heart-shape scarf inside a taxi.
He has no mercy
and for hours on end
crowned me "Miss Scarf of the Heart".
He just could not and would not leave me alone,
and then,
he saw the distress in my eyes,
and he stopped,
just as quickly as he started, and asked me softly,
"shall I take you to the Police Station to make a report?"

What an absurd proposition
and 'absurder' still
when he actually drove me there
leaving my bible study group dumbfounded
and his dear lovely wife speechless.

He stayed outside the Station as the sun slowly set,
to give me the privacy he thought I needed,
waiting in patience for someone in distress,
a distress he could not understand but accept,
and when I still did not come out after half an hour,
he got nervous and got inside,
in search of me.

Waiting in patience for half an hour
in the dark
for a silly scarf?
It must be too long even for the Saints of *Patience* –
but then who would have thought,
under his public persona of the tough macho man,
he is not only the very personification of *Patience*,
he is also one of *Discipline*,
Diligence,
Responsibility,
Righteousness,
Charity,
Generosity,
Traditions,
Sensitivity,
Kindness
and *Compassion*,
for the poor and the needy.
He would order a dish or two
when he's on his own
but a full table
with one dish too many
when he plays host to his friends lavishly.
He gives generously
to the orphans
whom he visits every week
and brings out for outings

at least once a month
not only presents and time
but paternal love and guidance.
He leads a special 'brother group in Christ'
where he is not afraid to get dirty.
He listens,
and he consoles.
He supports
and he encourages.
He confronts his own limitations
and when he makes mistakes,
as all men do,
he holds himself accountable.
He takes responsibility for himself
and is willing to be his brothers' keeper.
I once asked his grown-up son
what he thinks of his dad
and he said:
he has the kindest heart I have ever known –
and all this started with his fear of God,
this my dear friend,
and this fear of God alone
is the beginning of wisdom
and will be
to the very end.

November 2015

A Birthday Tribute to Jimmy
Our Dear Little Brother

There are graphic designers
and there are graphic designers –
and he is
without a doubt,
the one in italics.
Unique,
one of a kind,
a serious high-brow
who knows exactly what he wants to achieve;
he is the rare one
people like to work with,
for he has neither ego nor conceit,
but a readiness to rework his 'script'
and improvise with his 'clients'
whenever and wherever needed,
as he believes
that one plus one equals three,
if not more.
He is a highly esteemed member
of an elite Club
where membership entry
is neither by application nor networking,
but a hard earned approval
from fellow peers and end users.
I see how everyone is inevitably drawn
by his maverick vision
which turns every piece into art.
I know,
for I have seen him at work,
again and again,
over the years,
and marvelled every time,
at his amazing spirit,
where he has taken every opportunity
not just to set a standard

but to redefine it.
What is that spirit that has amazed us so?
They are passion,
inner strength,
ethics,
commitment,
imagination,
flexibility,
and the willingness to take risk.
And as I admire his artwork
one by one,
I am moved beyond words
at the rich body of his art
which represents
almost the full range of creativity.

There are graphic designers
and there are graphic designers –
but you don't see one
who still worked even at the very minute
before he's sent to the hospital
for an operation,
and not letting anybody know about it;
nor one who listened
with patience and indulgence
to every reasonable and unreasonable
demands made on him,
from the editorial team members
to ordinary classmates;
nor one who never complained
when the brief first given him
was simple and straightforward,
but as it evolved
gradually and steadily,
the implementation and execution

became so frightfully herculean;
no, you don't see any these days...
The distinctive look
of our 35th Anniversary Commemorative Class Book,
which has been wowed
by almost everyone who sees it
and dubbed the 'benchmark for all others to follow'
by our Vice-Chancellor
who officiated our 'book launch'
is the fruit
of his 6-month non-stop creativity and hard work,
the ultimate designer
who met every challenge
thrown at him
with sheer gusto and enthusiasm,
from designing a new cover
with every change of title for the Book,
to assembling
thousands of photos
of 103 classmates with families
and creating a total of 11 sections,
each with its own character
to complement the theme,
while retaining a uniform design for each.
His swift output of creative ideas
and his assiduous efforts
are all the more touching
as it is done for us
for free,
and for the love of his sister
who is one of us.

There are graphic designers
and there are graphic designers –
but few could remain friends

even if the partnership
has been smooth and satisfactory.
It's indeed quite different with us.
He is not just a friend now,
but our de facto little brother,
helping us with birthday party backdrops
where he once
dared to incur the very wrath of Sandro Botticelli
who must be turning in his grave
for the entire evening
to see how his Venus
unreservedly violated
with a new face and a new robe;
designing 'frames' to cradle our written poems
dedicated to close friends,
with one causing instant heartache
to a doctor friend
when he's presented
with an X-ray look-alike transparent copy
complete with the spine and all vital organs
with the poem on both sides,
and it's so hopelessly indistinguishable
from the real ones,
the kind-hearted birthday boy
rushed out quickly
in knitted brows
as he put on his medical hat
right in front of all his guests.
One is amazed
by all the exciting art pieces he has turned out
which are the result
of in-depth analysis,
thorough preparation
and tremendous hard work.
One does not doubt his artistic talent

he's born with,
but contrary to what one might think,
he has achieved success the old-fashioned way:
he earns it.
He does not just walk in and say:
this is my way!
He makes instead the meeting of the minds 'the way',
and it's sheer bliss to all who work alongside him.
He cares only about one thing:
to give his very best to every project,
no matter how small,
lowly remunerated or not at all.

There are graphic designers
and there are graphic designers –
yet it's the qualities
that he embodies in his life
outside his profession
that truly sets him apart.
The youngest son in a family of 12,
he is the very antithesis
of a spoiled, indulgent and frivolous character
one associates with the youngest of a big old family.
Ever so patient and attentive
and without losing his temper even once
when anybody would have
under certain unwarranted circumstances,
he remained calm and sweet-tempered,
and all these while he chose to stay at home
to nurse his dear sick wife.
His quiet and gentle demeanour,
his altruistic generosity,
and his natural understated manners
are made all the more impressive
with his fierce competitiveness

and enterprising resourcefulness
to render us only his best.
His soft 'wei' when he answered
my urgent calls
in the early hours of the morning
during our last month
before the deadline
of printing the Class Book,
had oft-times turned my eyes moist
with gratitude and humility.
And when I knew
he went in to Shenzhen
and had to stay for the night
just to get the right colour of the shells
he wanted for my birthday party,
I cried.
It speaks the richness of his soul
and how he has carried out his life
as a husband,
a father,
a son
a brother
and a friend.
Let's salute you,
our dear brother,
on your very big day:
may you and your family have health
and a joy that never ends.

May 2012

A Birthday Tribute to Fai
Our Very Dear Friend

He may be a medical doctor,
a specialist of the kidney,
a professor at medical schools,
a prolific scholar
with a stamp of high credibility
in all his articles and lectures,
a gate keeper
for quality control and risk management
of our public hospitals,
and like my very dear husband,
a high school chum of his,
who gives 200% of himself
to his chosen field,
he is a professional of the highest standard,
respected by his peers and colleagues,
but unlike my very dear husband,
who is all work and no play,
our dear friend
is the quintessential explorer
of the human experience –
the persistent seeker
of beauty,
simple or sophisticated,
common or rare,
transient or eternal,
whatever that tickles the senses,
refreshes the mind,
uplifts the spirit,
captivates the heart,
or soothes the soul,
eagerly sharing it
with his loved ones
and those around him,
delighting,
and indeed enriching

their very being
with his contagious love of life.
With a click of his fingers
or a tramp of his feet,
his world enfolds before us,
in digital language and telling visuals,
on swaying arms and twirling legs,
so richly colourful and enchantingly elegant –
that time stands still –
his slight hint of a smile
waiting ever so gently
for our exciting wow, wow, wow...
He is an inspiration.

He may be a medical doctor,
a one-of-a-kind physician
turned administrator
whose out-of-the-box thinking
and rare wisdom
makes him always ahead of the times,
yet his professional accomplishments
have never clouded his perspective
on the importance of life
outside of the hospital.
He cherishes family,
traditions,
culture,
art and music.
He is widely known for the parties he organizes,
but unlike my very dear husband,
also a veteran party organizer,
who only organizes political 'parties'
for votes and seats,
our doctor organizes social parties
for fun and fellowship,

to the joy and entertainment
of all his friends and loved ones.
He is the seasoned event organizer,
of every possible kind,
be it a wedding anniversary,
a surprise birthday party,
a charity fund-raising concert,
a New Year gala celebration,
a school alumni dinner,
or even a small casual gathering
which he would eventually turn
into something memorable for everyone.
His magic lies
intellectually in his brainy head
and physically in his little iPad
which he has surgically attached to him.
While we find the internet
huge, unwieldy and confusing
and easily get lost and tangled up in the web,
our doctor is the consummate IT expert
when it comes to the information superhighway
where he most happily
surf, design, develop and discover just about anything.
He's the programme and floor manager
of the surprise birthday party
my husband organized,
or rather 'decided' for me,
and it was planned and executed to the minute
with such surgical precision
it was absolutely timing-perfect
except for the unfortunate oversight
of the unperceived longer time
ladies required to spend in washrooms
during emotional moments.
I myself am also witness to his IT supremacy

as he turned 500 photos of a friend of mine
whom he has briefly met
and knowing neither one from another
from the sea of photos
into a 15-min photo video
interestingly grouped under ten categories,
each complete with a befitting music theme,
with not even one single inaccuracy.
He may be a medical doctor,
and an IT wizard extraordinaire,
he is also a world-class photographer,
a holiday aficionado,
a globe trotter and traveller,
a professional ballroom dancer,
a philanthropist and charity founder,
and above all,
the most devoted family man
and friend we know.
Like my very dear husband,
he is a filial son,
an adoring father,
and indeed a most loving husband,
with a wife
who is at once an outstanding professional,
a philosopher and a poet when her spirit moves,
a fashionable Armani girl
sporting also Chanel and Shiatzy Chen,
a soul mate of a friend,
generous and kind,
gentle and humorous,
and someone you can totally trust –
a role model of a wife, mother and friend,
the pride and blessing
not only to her husband and sons,
but to all who are fortunate to be her friends.

They have a unique parent-child relationship
with their two sons,
spending quality time with them
to familiarize them with all things under the sun,
sending them abroad for a western higher education
to widen their horizon,
and spending every holiday with them,
but unlike my very dear husband
who likes to take us to Shanghai, Beijing, Yunnan,
or other poorer parts of the Mainland,
they take them to all the exotic places
to experience the sun,
the snow,
the falls,
the boulders,
the forests,
the volcanoes,
and the deep blue sea...
chronicling their growing up
with post-card photos
and Hollywood-style videos;
so it's no surprise
that they would,
by practice and by habit,
stop whatever they are doing,
and come quickly to a complete standstill
with a ready-for-camera-smile,
at the slightest wave of his hand
or ripple of his brow.
Our dear doctor
is always ready,
steadfast,
helpful,
willing,
supportive,

understated,
in good humour,
and a good sport.
He opens his world
and shares it with his friends.
He is a massive character,
and the very incarnation of the perfect life balance
we teach our children to develop
but never quite achieve ourselves.
He is an inspiration.

Happy Birthday,
Fai,
we adore you.

February 2012

A Birthday Tribute to Alan
Our Dear Big Brother

He is soft,
he is tough;
he is easygoing,
he is amiable,
yet he could be meticulous,
even severe;
he is calm,
he is cool,
yet we have seen him
all flushed with emotions.
He is always cheerful in company
yet he has confessed
to being melancholy at times
when on his own.
He is the entrepreneur
who dares for literature,
and a literary highbrow who speaks
the language of the street.
He is the big macho tough guy,
our big brother,
and a sensitive,
thoughtful
and sometimes painfully vulnerable,
sentimental little sister.
While many of us emigrated
to Canada, Australia and the USA
during the 70s and early 80s,
he stayed not only
to embrace our return to the Motherland,
he was one of the very first
who went in to work
where life there and then
was not for the faint of heart.
He is a man of striking contrasts.

One has to know him well
to like him
and it took me
more than three decades to achieve both.
Our 35th Anniversary Commemorative Class Book
is our intermediary,
our little red string that ties us together,
like a comrade in arms,
fighting a 9-month war of words,
his in Chinese and mine English.
How he jealously guarded
his scholarly reputation
and how he spent many a sleepless night
to ensure only perfect writing and editing.
There's no compromise for him,
only the best.
He is all about teamwork,
and it's such a blessing
for an editor to be around a player
who has neither ego
nor a personal agenda,
only a winning attitude of 'I can and I will'.
Now he earns the admiration and respect
of his teammates
of those who have not the fortune
to know him well before.
He is a warrior of daily life,
of ordinary and normal things,
of the here and there;
a hero with no epic nor halo,
but a generous
and considerate lionheart
to all his friends.
He is a man with a big heart.

And so on this your very special day,
let me make a pledge to you –
let me be a sister to you forever:
let me compensate for all those long years
I have known and not knowing you.
Let me,
together with the G-5 members,
be your family forever.
Let us be with you when you in vain
still have not found someone by your side.
Let us be the stick
you would like to hang onto
when it is difficult for us to walk without help.
Let us be your ears
when you wish to share your joy or miseries.
Let's always put aside
at least five evenings every year
to go over and over again
all the trivial,
ridiculous and scandalous
events, incidents and episodes
we have gone through together,
as we sit at the Derby,
just the five of us,
a family,
for as long as we still have a breath.

Happy birthday,
dear brother,
may you have health
and a joy that never ends.

December 2011

A Birthday Tribute to Edith, My Dearest Friend

On that cool November morn
almost thirty years ago,
God came to my heavy heart
and bade me close my eyes.
"I have something for you," He said,
and gently laid something soft
in my hand.
In earnest
I opened my eyes
and saw a gift
elegantly wrapped up
in a most gentle smile –
it was you,
my friend,
it was you,
and I knew,
back then
how blessed I was
and indeed how blessed I have been
all these years
with you by my side,
in good times and bad...
For God must have known
that there would be times
when I need a word of cheer,
a triumph to be shared
and a tear to be brushed away.
He must have known
that I need to share the joy of 'little things'
in order to appreciate
the happiness life brings.
He must have known
that my heart
would sometimes throb with pain
at trials and misfortunes

or goals set too high to be attained.
He must have known
that I need the comfort
of an understanding heart
to give me strength and courage
to remain joyful
and be a blessing to others.
He must have known
that I need a friendship
that is lasting and true,
and so He sent you to me.

How people see you
as the super wonder woman that you are,
your never-ending stamina,
your unflinching determination,
your dedicated commitment,
your firm leadership,
your uncomplaining discipline,
your perceptive insight,
your vibrant energy,
your engaging personality,
but I see more,
by far:
I see joy,
tenderness,
sympathy,
mercy,
humility,
leniency,
consideration,
philanthropy,
kindness
and above all
compassion;

I see a reassuring smile,
a hidden tear,
a kind word,
a sympathetic pat;
I see a spring shower,
a dewy flower,
a song,
a dance,
a sonnet,
a star,
sunshine and moonlight;
I see a helping hand
to those who are weak and in need;
I see tenderness
and that angel touch
that tickles the heart
and soothes the saddened soul;
I see strength and the courage to be meek;
I see intuition and sensitivity to friends,
the thoughtful kindness
that turns doubt into faith,
drudgery into purpose
and disappointment into joy;
I see sacrifices made for friends
with a ringing phone
or a beeping Blackberry;
and a heart full of love
blessed by God's amazing grace.
I see the busiest woman
who always makes time for friends,
particularly the young ones,
who badly need direction and guidance.
Do you still remember
what you once said
when I could not sleep?

"Call me every half an hour,
I don't need to sleep you know..."
And when I sent you an SMS
at the late early hours that night
saying that I would not call to disturb you again,
you replied,
"thank you for being so considerate!"
My heart cried in joy
for a friendship sent only from Heaven.
And so on this your very special day,
I like to make a pledge to you:
I would like to be the sort of friend
that you have been to me;
I would like to be the help
that you have always been glad to be;
I like to wipe the grey
from out of your skies
and leave them only blue;
I would like to comfort you
with kind words
I so oft have heard from you,
and give you back the joy
that you have given me –
yet that's wishing you a want
I hope you will never need.
So just know in your heart
the sort of a friend
that I would like to be to you.
Let's do the small and big
and splendid things together;
let's travel on undaunted
in the brightest or darkest hours
with each other to lean on,
knowing deep in our heart
that we will remain friends

forever and ever...

Happy Birthday,
my dearest friend,
you have indeed been more than a friend!

October 2011

A Birthday Tribute to Cindy
Our Guardian Angel

She just sat there,
on the golden sand,
her halo tilted sideways,
her robe rumpled
her hair unkempt,
and her harp lay mute at her feet...
for months she just sat there,
sighing in despair,
unable to do a thing...
Our Lord the Merciful
stooped in His Pity
and very lovingly granted her wish:
go, my child, go –
be my messenger if you must.
In trembling hands
the sweet angel held her pass;
and in lightning speed
she doffed her celestial garments,
and without even waiting to lay them straight,
she bade adieu to St Peter
who stood by the Golden Gate.
Soft chorus from Heaven
chanted a fond farewell
while the imps glared up
as they pattered
on the fiery drums of Hell...

And so she sat there,
our angel from Heaven,
at the Lily Pond,
in T-shirt and jeans,
her long hair wavy on her shoulders,
her smile lingering on her lips,
as she sang along
with the stringed music from the guitar.

Ne'er was there such a maiden
with eyes so pure and clear,
a face so wholesome and fresh,
a demeanour so serene and modest
and a voice so gentle and sweet...
the doctor-to-be was smitten,
moonstruck,
and fell head over heels in love with her...
Together they started a family
of love, truth and beauty,
made complete with two children,
a girl and a boy,
both rare gifts from Heaven.
A loving wife,
a doting mother,
a filial daughter,
a caring sister,
an adoring aunt,
and a friend with a heart –
yes, she is the love of our lives.
And now as she sits amidst us,
her grown-up children the glory of God,
we all know where she's from,
for whatever she does,
she does it for others
and never for herself.
You will never find her at home
before her loved ones return in the evening,
for she's out where she's needed.
"For God commands His angel
to guard us in all our ways,
and with her hands she will support us,
lest we strike our feet against a stone".
So every day you will find her
soothing the needy,

calming the frightened,
uplifting the disheartened,
comforting the sick,
encouraging the despondent
or providing for those
who are deprived, denied or in need.
A blessing she is
to all who have the fortune
to cross her path
and double blessing
for all those who are her friends.
Yes, we are so blessed to be one of them.
I know,
for I have experienced
her love,
firsthand,
again,
again,
and again.

Happy Birthday,
our dearest C,
WE LOVE YOU!

September 2011

A Birthday Tribute to Elizabete
Our Dearest Comrade

We are the same breed,
yes,
Fung,
we are,
you and I,
not that we share the same surname,
or born in the same year,
or attend the same university,
or take the same majors
or wear only short hair....
I already knew,
back then,
when I first met you,
when we were but only 18,
that we bore,
and still bear,
the same stamp
on our very face
that represents
that can-do spirit of Hong Kong
that makes it flourish so....
How hard we tried,
on anything and everything –
from school,
to career,
to friendship,
to family,
both our own and the new ones we started...
how we persevered,
beaming,
upbeat,
optimistic,
always in good spirits,
persistent,
unafraid,

unshakable,
strong-headed,
stubborn,
single-minded,
striving only for the best,
remaining confident when pressurized,
unbeaten when challenged,
fearless when confronted,
and undefeated when discouraged...
and how young we were then,
when nothing was too difficult for us,
and only the sky our limit...

How you got out from your comfort zone of advertising
to work on pioneering projects
that you had never dreamt of,
and ventured finally
into mobile communications
in both home and virgin lands,
to become the magical classic Hutchison Wonder Woman
that I like to tease you on:
the most loyal employee
who never complains;
the most well organized executive
who works compulsively around the clock;
the most capable quicksilver
who multi-tasks
with two or three Blackberries
surgically attached to you –
yes, you are
dutiful,
dedicated,
diligent,
dependable,
disciplined

determined –
all top grade
for these all-important 6 Ds
any giant corporation
could only dream of having in their staff,
and you have them all.
It's no wonder
they will never let you retire,
and it's also no wonder
we will never let you go,
whether it's organizing our quarterly alumni gatherings,
or co-ordinating special social events,
or planning charity functions
or overseeing the completion
of our well acclaimed commemorative Class Book –
we simply can't do
without your legendary efficiency
and well celebrated high EQ.
No, nothing will be the same
without your participation
even if it sometimes means
you have to fly back just for us,
under the pretext
of the need to visit your hairdresser.

And so on this your very special day
when you are 18,
and 18
and 18
and a little more,
let me challenge you
to shed your skin
that I might shed mine as well...
Yes, let's let our hair down,
not that we have much to let down,

but down let's do,
and have some real fun,
not 'work' fun,
not 'being good' fun,
not 'taking care of others' fun,
not 'helping the needy' fun,
just plain
'lying down wasting time doing nothing' good fun.
Let's laugh aloud,
scream a little may be,
cry whenever and wherever we incline to,
throw a tantrum for no good reason,
behave childishly in front of our children,
get unreasonable for once,
and enjoying every bit of our nonsense
while we are at it...
Let's run in terrible glee through worlds
we have never been to
because we have been so tied up in the past;
let's lean on a lamppost,
if we could find one,
like a Hollywood star,
and smoke the cigarette
we have not even tried once;
let's just lie down
beside a bramble of red roses
and daydream,
no contemplation,
no worry,
no planning,
no reflection,
no thoughts even,
just daydreaming...
Let's loosen our rule books
we are holding on so tightly

all these years.
Let's go to the Paradise Hotel,
wherever it may be,
check into its Presidential Suite
and be the 'spoiled princesses' we never were...
Yes, let's just do all these
and have our shocked C taking photos of us,
to be taken out
to be enjoyed
every year,
as we sit
at the Derby,
at the same table,
just the three of us,
year after year,
until we turn 18
and 18
and 18
and 18
and 18 more...

Happy Birthday,
our dearest Fung,
WE LOVE YOU!

July 2011

For Eliza
My Dear Friend

"My regret
becomes an April violet,
and buds and blossoms like the rest."
　　Alfred Lord Tennyson

For three years,
it would come to haunt me...
on and off,
it would come,
when I am least prepared,
tormenting me,
disturbing me,
crippling me
with regrets
sick with misery and in low spirits...
When it comes,
I like to brush it aside,
as quickly as I could,
yet at the going down of the sun
and in the morning,
it will come back again,
to remind me
of my inadequacy,
an inexplicable incompetence,
bruising my pride,
mocking my self-esteem
and engulfing me with shame....
Yes, I am the most miserable comforter there is on earth...
We just had the lengthiest lunch ever
at the Country Club,
me and my good friend –
for hours on end
talking about everything and anything,
not forgetting,
for every hour or so,

to thank God
for having received so much in so many ways...
Two weeks later,
I received her call,
so strange was her voice
I could hardly recognize it.
"I am going to have an operation
for cancer tomorrow,"
she sobbed,
and when I finally understood her,
I could utter not a single word –
I just cried and cried and cried,
for more than three minutes I cried,
like I was possessed,
like I was the one who had the cancer,
like I was bestowed the precious 'gift of tears',
and my poor friend,
a woman of good fortune unacquainted with grief,
had to comfort me instead...
Of all the regrets that harass the distressed,
surely the most painful
is the inability to give comfort
when needed.
Nothing could explain
why I did what I did,
when I have always been
so compassionate with human sufferings,
so ready to give and help in whatever way I could,
so ready and easy with words...
and yet no words came,
no comfort given,
and no help rendered –
was I too stunned by the sudden change of fortune?
too shocked by the frailty of human life however privileged?
too terrified that it would come to me as well?

Like under a spell I was hardly myself –

When she saw me
as she began to slowly open her eyes
after the operation,
she held out her hand and said,
"you shouldn't have come,
the hospital is so far away."
She understood
and her kindness exonerated me.
So I too must myself set free,
for in Ecclesiastes, 4:21,
it says,
"and there is a shame
which is glory and grace."
I thank God
she is now very much alive and well,
stronger and unafraid,
embracing joy and grief.
And I too am fine
and doing well.

May 2010

For My Philippine Angel
The Ultimate Mother

She was forty
when she came,
a mother of five,
the youngest barely one,
the eldest but nine...
She came
not thinking of herself,
but her children,
that they might
have a better life...
For twenty years
she toils...
going home
once every two years,
with short trips home
for graduations or funerals in between...
"I want every of my child
to have a university education"
is the fire
that urges her on.

"Mam, my eldest daughter
has just graduated,
can you help her find an employer here?"
An employer here?
my heart sank...
a graduate of twenty-two,
so young and so full of promise,
a time
to dream dreams;
to aspire after self-realization;
to try out new things;
to indulge
may be once in a little while
her fantasy...

yet there's nothing
for her back home
and her only hope
was to follow in her mother's footstep
here in Hong Kong...
Now four years on,
the youngest also graduated.
She arrived here a few months ago...

"You are the best employer!"
she has always said.
Am I?
I am not sure,
I want to be
but I am not sure...
how maddening I could be
when I am working
on my poems –
papers flying,
notes here and there,
can never find anything,
always in desperation,
crying for help...
how many years
she has been
boiling herbal medicine
for me –
big fire,
medium fire,
small fire...
with every new doctor,
every new boiling method,
every new drinking schedule...
yet she remembers them all!
How many times a hungry stomach is fed

with a bowl of hot noodle
even in the middle of the night
when the angel
just suddenly appears...

All these for what?

Philippine women of generations
have come here,
leaving homes and their loves ones,
most having
to serve families
of three generations
with little babies
to cry them up in the middle of the night
and sleep in makeshift beds;
a few lucky ones
serve couples
with grown up children
and have their own rooms and toilets.
Some are treated like 'servants',
only very few
as members of the families.
On Sundays,
they used to congregate
at the Star Ferry and the Statue Square,
that's when they first came
some forty years ago,
now they move to
the flyway in Central and the Exchange Square,
still without a place of their own...
Both Governments close their eyes,
devour their conscience,
if they have any,
one getting all the domestic help

the other the foreign currencies.
The girls could rot –
for decades,
still no site has been designated
to provide resting and recreational facilities
for these angels.

Mine is
the best any employer could have,
a most valuable member
of the family –
but is that enough for her?
You may try provide her
with everything possible,
but something
deep inside her
must be hurting –
just by being here
year after year,
year after year,
with all her children
growing up
without her
by their side...

If there's any consolation
for her,
it is,
after twenty years,
she is able
to see her daughters
every week.

May 2008

For Edith
Music of My Night

When she said yes,
my eyes turned red,
my paper tiger,
my dearest friend –
the busiest globe-trotting executive,
the eternal giver;
and the perfectionist,
who speaks perfect Putonghua,
would fly in to Beijing
to speak about me and my poems
at my book launch...
"Could you sing for me as well?"
I was so greedy.
"Yes, I will.
Can you find me a piano?"
A piano?
I can find her the moon!
When time finally came
for her to go up the stage,
the chatting stopped,
there she stood,
poised and elegant,
the last to speak,
in her charming cheongsam...
in earnest I listened and waited for her songs,
and when they came,
the audience was mesmerized....
they won't let her go –
"I thought she is a lawyer!"
eager questions from bewitched ears.
"The Three Wishes of the Rose" she sang for me,
I was the rose,
she said,
without the thorns...
Like under a spell,

we thirsted for more,
she obliged,
and gave us another
amid thundering applauses....
A grand finale she gave me,
to end my special moment –
she was the Music of my Night.

The grand piano was moved to the China Club.
The pianist did not seem to care for dinner,
she sat at the piano for the entire evening,
her fingers flying over the keys,
they never stopped...
How we sang,
singers or no singers,
how we sang,
how we enjoyed ourselves,
like we were on fire...
Even our guest of honour,
the Exco Convenor,
shed his public persona
and sang like one of us...
A lawyer friend from Beijing
stood on his crutches,
and sang songs we have never heard before,
how my pianist could manage
we neither wished to know or bothered to ask...
Then the duet came,
from my diva,
"That's All I Ask of You"
closing the curtain of a special night.
Another grand finale she gave me...,
she's the Music of My Night!

April 2008

For Ha Ying
She Stole My Heart

She sent me a poem,
her very first,
capturing a moment,
my special moment,
on stage...
she was down there,
her eyes on me,
taking in everything,
not just a snapshot
of my physical being,
nor the environment,
not just my 'stilettos',
'horn-rimmed glasses',
nor my 'pearl' strand,
not just 'the mountains of flowers',
nor the 'dignitaries'...
not just the way I stood,
the way I carried myself,
the voice I spoke:
she sensed my anxiety –
how could she?
Did she see my clenched hands or shaking legs,
feel my clammy hands
or hear my quickened heart beat?
I was vulnerable,
hardly myself,
to be in tears any time soon,
and she saw it all...
so vividly described,
and so tenderly rendered –
I read it again and again,
I even read it to myself,
aloud,
my first one from any other,
apart from him,

last received twenty-five years ago...
She stole my heart.

The Beijing traffic was irredeemable –
For an hour I sat in the unmoving car,
so desperate I jumped out of it
and started to run in my stilettos,
for half an hour I ran,
how I ran and ran in my appalling stilettos,
my hair flying,
my heart beating fast,
not knowing when or how one of my earrings
dropped from my unpierced earlobe...
Arrived finally,
breathless and totally ruffled,
with not a minute even to comb my hair.
Guests were already filing in...
hello here, thank you there,
"how kind of you to come",
with my husband helping me along....
Then I was on stage,
my mind still totally blank...
I stood there,
opened my mouth and no words came...
I looked at the staring eyes below,
waiting in anticipation,
yet no words would come,
my legs began to shake,
but still no words came....
for a lifetime no words came,
then I saw her,
comforting me with her smiles,
supporting me with her ardour,
and when our eyes met,
like a prompter she was,

she gave me my words and I was myself again –
words came flowing until the end,
nonstop,
with applauses in between...
She stole my heart.

Another moment,
she too captured,
a moment of celebration,
at dinner for all friends
who flew in from Hong Kong
for my poetry book launch.
The grand piano
temptingly stood there,
inviting fingers to dance on it.
How our friends sang,
one after another,
singing together,
with a few invited to sing solo
while the more expressive danced to the music...
then she was on stage,
urged by me,
to read a poem
about an evening we spent together,
how delightfully she read,
everybody was listening,
in silence,
soft but clear her words came,
at times fast,
at times leisurely;
at times forceful,
at times so soft,
we all leaned forward to hear
lest we might miss a word...
in deliberate rhythms and poetic accent

she read,
a dramatic pause here,
an assertive word there...
breathing life into my poem,
a long one but too short for our ears.
Such an enchanting delivery,
like poetry recitation is her profession.
She stole our hearts.

A medical doctor by profession,
wife of my husband's high-school chum,
another medical doctor,
a most adorable couple...
"You should tell them about your book launch in Beijing"
my husband insisted.
"I won't, it's too much of an imposition."
"To inform, not to impose," he was adamant.
A phone call and they were on their way to Beijing
when their son was to fly in from the UK
on that very night...
Yes, they just flew in to share with us
this our very special occasion,
making it even more special for us.
They stole our hearts.

March 2008

For Stephen at 60
Our Understatement

"Nobody loves a fairy when she is forty",
not her,
even when she turned two and forty –
energetic,
dynamic and determined,
our fair lady is a tower of strength –
in London one day
and in Milan the next,
and straight back to Hong Kong
without a break...
her poor Blackberry,
surgically attached to her,
works like a digital dog,
without time
even for an electronic bone...
a lawyer by profession,
a performing singer by DNA,
a Columbia alumnus extraordinaire,
a trained educationist,
a loving daughter,
a caring sister,
an adoring auntie,
a devout Christian,
and a friend with a heart....
Strangers may be
intimidated by her stature,
but who would not be?
Her air of authority
exudes naturally
in the way she talks,
in the way she walks,
and in the way she carries herself...
yet when you get to know her,
she is but a paper tiger...
She may have no time to chat,

but you will receive her emails
sharing thought-provoking ideas,
funny anecdotes
and encouraging messages
when you need them most...
And for those who are fortunate
to be her friends,
there is always time to weep,
time to laugh,
time to mourn
and time to sing and dance...
Yes, she is a paper tiger
with a heart of gold.
I have experienced
her kindness first-hand,
again and again:
"Call me whenever you have to,
I don't need to sleep you know"
she once told me
when I was in distress,
and all she wants
on her tombstone is
"She has been a Friend"....
Perfection is
but a jealous mistress
and Fate played her
a little waiting game....
our fair lady had everything
but her Prince Charming
was yet to come...
No, it's not that God
wanted to let her wait –
"but who is good enough
for our prima donna?"
we could often hear Him sigh....

and so for years she waited
and waited –
Yes, God let her wait
and then
gave her the best....

When our Prince Charming finally appeared,
God made them a couple in seven days....

Fun loving,
sociable,
a baritone,
a keen pilot,
a seasoned scuba diver,
a wine connoisseur,
our prince is
a 21st Century Renaissance Man –
polished, subtle and reserve,
he underplays his talents
and laughs at himself,
at every possible opportunity,
earning respect and friendship
from both his friends and business enemies –
an understatement
he is,
and that's the way he wants it to be.
How he wooed her with his charm,
comforted her with his love,
moved her with his songs
and conquered her with his heart!
Never had more diverse personalities
so complementarily joined.
They complete each other.

And so,

Adam and Eve
live happily ever after,
in the Garden of Eden
where flowers bloom in all seasons,
where trees bear fruits
of every kind
except apples,
where dogs and cats are good friends,
where serpents hibernate 365 days a year;
yes, this is not only a land of milk and honey,
this is a land of red wine
and oysters and oysters and oysters....

And so on this your very special day,
let our soul clap its hands together
and sing...
let this room
in every corner
sing praises to the Lord;
let's celebrate,
hopefully with approval from our lady,
with red wine and oysters,
and more red wine and oysters
until we all fall to the ground....

Happy 60th birthday,
Stephen,
WE LOVE YOU

January 2008

A Birthday Tribute to Virginia
The Queen of Words

"Nobody loves a fairy when she is forty",
not her even when she turns fifty –
for like Cleopatra,
"age cannot wither her,
nor custom
stale her infinite variety" –
A fury slinging flame
she is,
burning us with
a fire so scathing,
a power so savage,
a vigour so fervent,
the fainthearted
may wish to meet up with her
no more
than once
every two months,
for fear of being scorched,
a literal condemnation:
high blood pressure,
stomach ache may be,
and for those lacking in self-restraint,
minor stroke even,
(too much uncontrollable laughs
could be fatal if not disastrous),
these and other malaise
that 'dares not speak its name',
lasting for at least two days and a half
and not easily recoverable
either....

For our heroine
is the queen of words:
she lives and breathes amid a wealth of them –
they are her weapons at work,

her magic wand at play,
and her fun gadgets at home.
Her words are her chameleons,
reflecting the colour of her thoughts:
they come in different shades,
in different languages and dialects,
in duplets or triplets
or in pairs of four,
with multiple meanings
intertwined in her delivery:
at times leisurely,
at times fast;
at times piercing,
and at times so soft,
everyone bends forward to hear,
lest a vital detail
might go amiss!
Graceful and eloquent
her words come,
as if on cue,
never a second too early
nor a second too late...
in deliberate rhythms
and poetic accent,
she tells her tale...
a dramatic pause here,
a loaded gesture there;
an outrageous observation every now and then,
and delicate subtlety throughout...
ne'er a tale is more vividly told,
even when it has already been talked about
over a hundred times:
there is always
one secret more to give away,
a scandalous infamy to uncloak,

a shocking impropriety to divulge,
a juicy affair to reveal...
all these
to quench
our insatiable thirst
for intellectual development,
academic curiosity
and simple and pure
noble joy!

Now you know why
we so much look forward
to our luncheon party,
eagerly awaiting
to be amused,
entertained,
shocked,
bewitched,
by events big or small,
trivial or profound,
ugly or beautiful,
honorable or disgraceful...
listening to all these
while feasting our eyes
on her animated face
and her expressive eyes,
shining
with such pure ecstasy,
radiating a sunshine in any shady place...
So what is there
to have a luncheon gathering,
a field trip,
mountain trekking,
an afternoon tea,
or a shopping spree

when she is not there?
Yes, she may be a windbag to some,
and she could be merciless,
she atones for her occasional overkill
by being always so refreshingly delightful,
so utterly convincing,
and so intoxicatingly quick-witted,
one is touched
by her fiery sensitivity
and her conscientious dedication
to perform her 'duty' to perfection,
never allowing us to forget
Wilde's famous epigram:
"moderation is a fatal thing,
nothing succeeds like excess"

Our lady lives her life
to the fullest
and much more,
"while most of us
only stay with it
or at best
lunch with it"...
So let's celebrate 'this work of art',
let's thank her for those exquisite moments,
let's crown her with laurels
on this her very special day,
let's fill up her scrapbook
initiated by her most adoring daughter,
(what sigh of relief for mothers who have only sons),
let's recapture all those fun and laughter we have shared
together,
let's tell her she is not just a super mum,
she is indeed our super gal,
fifteen or fifty,

she is all the same to us
and we adore her.

All these acclaim of her genius
I now shall stop,
organize a dinner,
and just be there to hear her speak –
whether her witticism
is diligently prepared
or studiously rehearsed beforehand
is for you to judge -
I hereby rest my case.

June 2007

To My Doctor Friend
The Innocent Fish Bone

"Excuse me," I murmured.
Calmly I walked to the washroom,
slowly, ladylike –
then into the toilet I dashed,
trying to cough out
my little agony –
yet no amount of spitting,
wheezing or clearing of my throat
could do the trick.
I tried everything,
everything possible,
again and again,
but the little monster was in hiding,
refusing to come out!
The washroom lady was concerned –
knock, knock, knock,
"do you want a glass of water?"
as if I could still answer her
while my throat was rasping,
my face a roasted pig red,
my tears running
and my mouth drooling...
then the lady captain came,
unsummoned,
with a glass of warm water,
then the manager,
with a bowl of rice,
a glass of vinegar
and a plate of sticky malt candy,
explaining how each could work:
my eyes surveyed the remedies,
my mind a total blur,
and just as my fingers
switched from one to another for the magic cure –
"no, none of these!"

a voice from somewhere,
so gentle and soft,
yet so full of authority.
My fingers stopped in mid-air
and my dazed eyes saw the image of an angel,
yes, my angel,
my dinner-mate,
a senior lady doctor –
yes, none of these,
she could simply take it out for me!
"Go see an ENT doctor now," she declared.
At this hour?
Yes.
Frustrated and too exhausted to argue,
like a little lamb
to be sent to the slaughterhouse,
I let my angel take me to the hospital,
my splitting head
haunted with the harsh gentle words
of my angel's husband,
head of a cluster of hospitals:
"it's nothing–
the doctor will anesthetize
your palate and posterior pharynx with a spray
and he will then remove your agony
with a pair of forceps.
A mirror is all he needs."
The 3-minute ride to the hospital never ended,
not when you were being explained
on what would happen if unlikely complications
were to occur...
The doctor was too young for anybody's comfort:
after being fully explained on what happened,
he asked me to sit down which I refused to save time,
so with a little torch and a tongue depressor,

he began stirring,
shifting,
whisking,
prodding
the base of my tongue,
my tonsils
and anywhere he saw fit....
No, he could not find it,
again and again he tried,
but again and again he failed...
he then asked me to raise my palate
by panting like a dog,
and for the first time panic seized me:
from what he was heading,
he might easily push it further down my airway,
cut my esophagus
or put a hole in my intestine...
I could see a parade of physicians, surgeons, ultrasound, CT scans...
a sudden spasm gripped me
just as my angel was about to lecture him
on what he ought to do –
I vomited everything
I had in my stomach
including my little 3-cm demon
which had tortured me for the past 30 minutes!!
Gone just like that,
suddenly,
quickly,
almost at the speed of light –
gone was the pain
like it had never existed.
The young doctor
continued to ask me questions
but I heard nothing:
I fled like the wind gasping for air:

I wanted to savour my joy untrammelled
by any questions,
any enquiries,
any noise,
any sound...
When we returned to the Club,
I was glad the whole fish was finished:
they were advised to have it when it was still hot:
who could have guessed we would return so quickly!

Now when I pray at night,
I pray for something more:
to guard my soul
against sins of the tongue and of the palate:
they are areas more dangerous than any
that can afflict our body and soul.
Amen!

March 2006

For GN
A 'High' Ordeal

I thought I could never survive it –
at one point
I collapsed,
and cried for help,
real help,
a helicopter,
an angel if possible,
to lift me up
from this hell of misery:
but the terrifying image of me
being hauled up
in the news the next day,
my body swaying in the wind,
my hands desperately grasping the rope of hope,
failing again and again
to get onto the helicopter –
on every news spot
of every channel,
and on Cable
once every half an hour,
with the identities
of the two men
unscrupulously flaunted
with the minutest details
calmed me down...
With my husband's hand
gripping mine on my right,
murmuring words of comfort,
and the 'gentleman's' on my left,
kindly urging me for calm,
I plucked up my courage,
and stood up again...

But the nausea would come,
again and again...

"An afternoon walk?"
"Now? It was storming this morning!"
"It's not raining now!"
I rose to the challenge at once –
I am a keen walker,
I could walk for hours and hours,
the entire day
easily...
"Let's start at 4 pm from my place,"
our gentle friend said.
We rushed home
to change into our sneakers and T-shirts,
not forgetting our walking gear –
the bottles of water.
When we were there
he was all set and ready,
putting on his track shoes
and his backpack –
I had to hide my smile –
here's a gentleman of the old school
getting too equipped for a walk,
but while he knew
what laid ahead
he did not know
that
I did not know...

"We'll start from these steps up,"
our friend of class
pointed at
the three stretches of steps
in front of us,
one steeper than the other!
"I have serious height phobia"
I wailed in desperation...

unflustered and composed as ever,
our English public school boy comforted me:
"It's flat all through after that,
and you don't ever have to walk down."
Much encouraged,
I willed myself to be calm;
in less than fifteen minutes
I made it to the top,
never stopped for a second
to look around at the scenery,
nor down to test myself...
As I breathed in relief
waiting for the two to catch up,
I suddenly realized
what I was getting into,
and I froze...

I 'froze' up for the rest of the 'walk'....

The muddy track was indeed flat,
slightly undulated with stones
and moulds filled with rain water,
but only three feet away
(five feet to my husband)
(eight feet to our friend)
naked from the side of
the cliff
hundreds of feet
above the valley below...
with an occasional railing
for lovers to lean onto
for a bird's eye view.
I sweated profusely,
my heart began to beat irregularly,
I was on the verge

of a breakdown ...
Even the hateful mosquitoes
did not spare me,
my arms laid limp,
too worn out to flap them away
as they danced all over me
in steps of quick succession,
or one step at a time,
on my sweating face,
my drooping arms,
my stiffened legs,
my aching neck...
happily, hungrily, biting away...
I walked furthest
towards the rock,
rugged and bumpy
it might be
but much,
much safer in my mind...

With every step I made forward,
every 'good',
"you are doing well",
'excellent'
I received,
said gently,
kindly
patiently
almost patronizingly,
but in the kindest possible way...
he never stopped
those words of encouragement
until I made it
finally,
an ordeal

that lasted for
two hours...

To this unholiest terrain
he has innocently brought us,
with not the slightest idea
of what laid ahead for me.
But how could he know?
We are but social friends
meeting on social occasions
wearing a social veneer
demanded by polite society.
He gave me an ordeal
I could hardly endure,
but also something else
I could hardly forget:
a gallantry of years gone by –
for someone who laughs
and jokes at everything,
he has shown me
understanding and patience,
sensitivity and compassion
in my hours of distress...
For over twenty years
of knowing each other
but not knowing each other,
we are now embarking
on a journey
of knowing each other....

My husband and I look forward
to our 'first' dinner
on July 3.

No surprises, please...

June 2005

And the song, from beginning to end,
I found again in the heart of a friend.
    H. W. Longfellow

She came back early,
it was barely 10 pm.
I was surprised,
"did you not enjoy your dinner?"
"Yes, but I want to come home
to finish my bath and hair wash
before 11 pm"
Oh the wretched '11',
more despairing than that of Cinderella's...
To come home so early
when she was having dinner
with her longtime friend,
wife of the American Ambassador to China,
when fine dining and wining
until the small hours
is an everyday affair
for her....
She came out in her nightgown,
her feet in thin socks cold on the ground,
I bade her good-night.
I trusted not myself to say another word...
In my own room I lay down,
too touched to think,
and greatly disturbed
by the noiselessness –
not even the flickering of magazine,
book or newspaper pages...
how could my poor night owl
survive her next hours
until it's her normal time
to go to bed...

she did not even go to the bathroom!
Guilt gripped me –
like a demon
it tormented me...
I stirred and shifted,
but sleep would not come...
I reached for my sleeping pills,
and the good angel of sleep
took pity on me
and held me in her arms,
comforting me
like I was a little baby....

We took this service apartment
with two bedrooms
at the China World,
the first time
during all these years
when we went on for business.
She let me have the master bedroom
to discover only too late
that the guest bathroom
was right next to my room,
no, right next to my bed,
much closer to my own en-suite facilities.
The architect be hanged!
It was hell...
for three long wakeful hours
I could hear every of her movement:
flap, flap, flap,
in her slippers my night owl
was in and out of the bathroom,
click, click, click,
the opening and closing of the door,
saa, saa, saa,

the flushing of the toilet,
chue, chue, chue,
the running water,
and when she began to take her shower,
I knew the night was going to be long,
and I gave up –
Cinderella had long past her time,
and no amount of additional sleeping pills
could do the trick...
in resignation
I waited
and waited
and waited,
and dawn never came...
She knew at once what happened
when she saw me the next morning,
and for the next three nights
her suffering must be greater than mine...

Oh the pious friendship of the female sex,
sighs William Congreve in his grave –
my night owl
is not just a business partner,
she is a soulmate,
a confidante,
a sister and a mother
all in one:
"Thy friendship oft has made my heart to ache:
Do be my enemy – for friendship's sake."
Love is divine,
but the yoke too heavy to bear...
To William Blake
I solemnly make this pledge:
to separate hotel rooms and only separate hotel rooms shall
we next go,

I don't want any other way.

October 1993

A Garland of
Cherished Friendship

For the Suns
Exemplary Parents

And after much deliberation
and weighing up,
the loving parents,
with pangs of heartache
and a distress so miserably suppressed,
sent their two sons
back to school in the US again,
having to undergo
a three-week quarantine
this time
instead of the two the year before,
as the Covid situation in the US
has hardly improved.
I wanted to know why:
"We trust the school
with their strict social distancing rules",
so the Suns told me.
"It's a tough decision
and we want to do
what we think is best for our children."
But online learning is the new norm,
I persisted,
so why take the risk?
School education goes beyond academic
to include interaction with other students
and faculty members,
so the parents explained,
and with only online learning,
children are missing out
on an important phase
of their learning journey
and overall development.
The Suns are both elites
with a can-do spirit,
positive but cautious,

and put the boys ahead of everything else,
and if they believe
that's the best for them,
it must be the best.
And indeed the Suns are exemplary parents
who would do anything
to provide nothing
but the very best for their boys.

If one looks at the four of them,
and sees how they interact
with and among each other,
one might be forgiven
for thinking that good parenting comes naturally
at least for them,
when it is not.
Like both their own parents
who have been their role models
and successfully brought them up
to be such outstanding parents that they are,
our two good friends,
privileged in every way,
have still taken a long time
to learn how to be good parents,
having been tested on every possible level:
be it physical, emotional or financial;
and that they are still making mistakes
and learning every day.
To them a healthy sense of humour is top priority.
It's the same thing with common sense
only moving at different speeds.
A sense of humour
is just common sense but dancing.
Children must learn when young
that humour is one of the best components of survival

as it encourages open dialogue
and provides relief from stresses
that life inevitably brings.
It particularly makes family time so much more fun.
Our witty Mr. Sun is a scream,
a blazing flame he is,
so full of energy and life,
a dazzling sun,
and the best story teller there is,
being jested at for his persuasive artistry
that could trick down even birds from trees,
has successfully passed onto his two boys,
a spirit of playfulness and fun,
and it's pure delight to see them together
on various occasions,
with Mrs. Sun smiling her gentle smile
and playing along.

Open-minded and easy going,
they may respect their sons' rights
to have a voice in family decisions
from choosing their own friends
to the schools they wish to attend,
mediocrity in academic work is however not tolerated.
Regardless of how busy or tired they are,
they always find time for the boys.
Tutoring English, mathematics and science from him,
and arts subjects from her;
but nothing is more important
than bonding with them whenever it's possible –
watching baseball games and playing golf with the elder;
or playing games at the arcades
and miniature golf with the younger
are routines they do together all through the years,
developing a unique parent-child relationship

that has always been as solid as gold.
Integrity, humanity and generosity
are what they wish to impart to them
which they teach by examples.
We have known Helen for almost forty years
and she's like a sister to us;
We also know her parents well
and recognize where her sense of commitment comes from.
Then Leland came to her life,
and then ours,
and how he lives up to his name:
for like the 'Sun God' he is,
he touches us
with his enormous wealth of knowledge;
his keen interest in sports and leisure pursuits;
his ardent enthusiasm to live life;
and his total dedication to his two sons
who are now young men of such promise,
having been taught since young
to embrace challenges and persevere with every fall,
to create and innovate,
to think out of the box,
and be a leader rather than a follower.
No words could express the power and beauty,
heroism and majesty of their unconditional love.
My warm salute to our inspirational parents.

April 2021

Ode to My Women Friends

I learned of the magic of women friends
when I was still in my teens.
I grew up in a convent girl school
and had met,
nay,
experienced
almost every type,
in its purest and natural form.
On one side of the extreme,
there were holy angels
who lifted you up,
and on the other,
unholy devils,
with their little demon –
accidentally or by design,
put into and reside in them,
that no quiver nor cries
of the ones being bullied could remove –
literally pushed you down to hell.
Then there were
the shy and the bookish
who kept to themselves.
There were those
who day-dreamed constantly,
and there were
the enthusiastic and sociable
who had a knack for being friends
with almost every girl and social group,
the popular ones
in our young naïve mind.
There were the rare few
who got straight A's every time
and apparently without even trying.
There were the ones who knew all the rules
and followed every one of them,

while there were those
born with a headstrong attitude
towards authority
and were always trying
to prove the teachers wrong.
There were the inevitable
busybodies and gossipmongers
and those who got jealous easily.
There were the artistic,
the athletic,
the musical,
the writer-to-be,
who got invited everywhere.
Then there were those
whom no one really spent time with
and being unintentionally ignored
and that's the majority –
who remained quiet and mild,
uninvolved and apathetic,
silent and private,
indifferent to what's going on around them,
and thereby quite forgotten,
along with their names,
as one takes up the annual class book
decades later.

I was more aware
of the magic of women friends
when I became a mother.
There's so much
I didn't know about child care
despite being a Child Psychology major,
that I don't think
I could survive
my first 6 months

without the active help and support
of friends who were mothers
with babies and young children,
and I have been enlightened ever since.
So while most of my peers were busy
developing and building their careers,
starting and growing their families,
or social networking
with the rich and famous
with little or no time
for true friendship,
I spent time
making and nurturing it:
reconnecting the old
and discovering the new,
choosing carefully,
not just anyone
who happened to come by
but those whom we could trust
with even our darkest secrets
and share
our inner self without fear,
accept us for who we are,
and enjoy and appreciate our company
that we might become
stronger and happier
by having each other.
I have long realized
by then
how the right friends
could truly be life altering.
I remember how I started off
with questions
of what I wanted from friends
and what I in return

could offer them.
And how hard I tried
to better myself
to be worthy of them
and add value to their lives.
For while friendship might begin
by accident or choice,
no friendship
could pass the test of time
without persistent nourishment,
mutual respect,
shared values,
interactive give-and-take,
reciprocal appreciation
and a genuine desire of both
to treasure and cherish it.

So now I sit,
deep into the night,
writing a tribute
to celebrate
the magic of women friends.
And with this little poem,
my dear friends,
I want to say thank you
to each and every one of you
who have entered and touched my life
in your own unique way.
I want to thank you all
who have listened without judgment,
spoken without prejudice,
understood without pretension,
helped me without entitlement
and loved me without condition.
I thank you for inspiring me

to reach for the stars;
motivating me
to reset my focus when I fail;
encouraging me
to believe that I am capable
of achieving far more than I could believe.
Thank you for showing me
how to stay calm in difficult situations,
comforting me when I am down,
invigorating me when I fall,
exploring alternative options when I suffer defeat,
and applauding me when I succeed.
I am amazed at the depth
of the many friendships
I have been able
to nurture and maintain
and for that I am eternally grateful.
Women help each other
in ways big and small,
every day,
without thinking,
and that's what keeps us going
even when the world continues to come up
with unnerving ways to crush us.
When we band together
to stay strong and positive,
helpful and supportive;
to build and inspire,
to listen and nurture,
to laugh and have fun together,
it's like a beautiful firework display
mesmerizing us
with those bright breathtaking moving sparks,
spectacular colours of the rainbow
and the unpredictable dazzling gap

between the bewitching flashes
and the explosive snaps, crackles, pops...
a hypnotic cocktail of colour and sound –
it's a song,
it's a dance;
it's gripping poetry,
it's pure magic.

January 2021

Happy Pearl Anniversary to Christine and HF
An Exemplary Couple

When he met her
some thirty years ago,
the barrister-to-be was smitten,
moonstruck,
and fell head over heels
in love with her...
Ne'er was there a maiden
with a face so wholesome and fresh,
a voice so sweet and melodious,
a demeanour so gentle and yet so alive,
a mind so bright and original,
and so witty too,
that even the dullest man
would laugh aloud
when she's around,
but it's her big twinkling eyes that got him –
he could not stop looking at them –
for they could speak,
and smile too,
that any man would fall for them,
and he did,
in utter surrender,
irresistibly, wholeheartedly, completely...
never mind if she does not care
for cooking nor housework of any kind,
as she later confessed.
His fate was sealed.

On that warm Spring morning in early March,
thirty years ago,
God came to her trembling heart,
and bade her close her eyes.
"I have something for you", He said,
and gently laid something soft
in her hand.

She opened her eyes,
and saw,
in her laced left hand,
the warm loving fingers of a man,
gently putting a diamond ring
on her fourth finger,
the man she has loved
for four long years...
and she knew,
back then,
how blessed she was,
and indeed how blessed she has been,
to be bestowed a gift
so rare and precious.
For the Almighty God,
has long known,
that this man and woman
are made for each other,
to love and cherish,
until the end of time...

And how they love and cherish each other –
Good givers they are,
and good forgivers too,
they support each other
but allow for change and growth
as individuals
to bring out the best in them.
He is rock solid and dutiful,
and she knows
she could forever rely on him
for encouragement and support.
Her heart is tender but strong,
delicate yet resilient,
and she gives it to him,

her most prized possession,
which he loves, nurtures, cherishes and protects.
And like a good mediator that they both are,
they are effective communicators
and active listeners too,
they frame their words
before they are out,
and when they are wrong,
they admit it,
and when they are right,
they shut up –
so one does not need to wonder why
they have never quarrelled
for all these years going through life together,
except once
when they were still young, simple and naïve,
which ended
in Greek comedy...
That's a cold winter evening,
when they had their first,
and doubtlessly their last,
a quarrel
over something of principle,
that neither could remember what it was:
and she,
in a fit of anguish,
took her flight from their love nest,
and dashed down to the lobby of
their Hang Fa Chuen home,
expecting her man to follow
and beg her to return,
yet he didn't
and so wait she did,
in earnest,
but he never came...

and then
with gas in her stomach,
freezing air in her lungs,
hot smoke in her head,
and no mobile phone in her shaking hand
to get advice from trusted friends,
she made her 'Elopement Plan' –
whereto and with whom.
Yet before inspiration came
she heard the rumbling sound
of her empty stomach,
and for a life time of 10 or 15 minutes
she fought,
and she fought hard and well,
but how could a young person
triumph over the painful crave for food?
Our heroine,
otherwise of iron-will,
resigned.
Up she went back to her apartment,
but with her door key forgotten,
she was forced to ring the door bell...
she rang,
and she rang,
but no one answered,
and when he eventually did,
he was half-asleep,
and not knowing that she has 'eloped',
asked drowsily,
"why are you outside?
Are you not sleeping next to me?"
Our young lady learned her lesson
hard but fast,
and she never repeats her 'mistake'.

God joined two extraordinary people in holy union,
and blessed them with the pearl of a girl
who is a bundle of joy –
enriching their lives and that of their loved ones,
in thousand and one ways.
It's your Pearl Anniversary,
and we admire the good role model
your marriage provides for us:
you are a golden couple
who shines like a beacon of hope
for all married ones:
let's learn to glow as you do,
going through life
with such zest, joy and love,
sprinkled with wit,
and a good sense of humour.
Come everyone,
let's give a big cheer,
in celebration of a couple dear,
the ultimate couple who is love itself.
May you laugh, sing and dance
as you further sail through life's journey
with joy surrounding you three –
and that's what we wish for you.
Happy Pearl Anniversary,
Christine and Hing Fung.
We love you!

January 2020

Ode to St Mary's

We gather here tonight
to remember those our younger days,
when youth was a symphony,
and every day a song;
when life was such a promise,
and adulthood years away.
There were such fun and laughter
as we strolled the playground,
in pairs or in groups,
making serious plans for lunch break
or a Saturday movie
or a Christmas party,
with hidden romantic yearnings
for the "Prince on the White Horse".
There were so many learning activities –
camps, experiments and debates.
There's the most awaited
Joint-School Sports Day
with the other two Canossian Colleges,
with our sportswomen
winning more trophies
and our cheerleaders
chanting the loudest,
backed by a continuous
"SMCC" mantra from us,
the ardent spectators.
There's the unforgettable
"The Sound of Music"
our first school movie out...
There's a gossip here
and there's a bullying there;
and an occasional crisis
of 2 or 3 inches uniform length too short
or fancy hairdos
that we had to work through

with the help of our class mistress
to explain
to our otherwise most indulgent 'Mothers'
now renamed 'Sisters'.
But nothing is more important
that would preoccupy our young minds
for well over a month:
the Annual Drama Competition
complete with full costumes
stored somewhere
in that huge,
untouchable room...
Then June came,
all such activities stopped –
our carefree faces turned anxious,
everyone was hurrying,
running from one class to another,
from the classroom to the bathroom
from the tuck shop to a quiet corner
with books in hand,
'hammering away',
borrowing notes
from the brainy,
the good and the unselfish
all for a last minute revision
and all for a good score
where our future lied...

Most of us gathering here today,
have children fully grown,
and may be even grandchildren
to call our very own.
The careers we spent our life at
are now almost through...
It's easy for us to see

how different we all look today,
and how the rigours of life
have over the years left their marks,
but our heart remains
young and alive,
our mind
more seasoned and less judgmental,
and our inner being,
much wiser,
more loving and forgiving...
Let's take this special moment,
to thank each and every of our teachers:
who have planted
the seeds
of curiosity and motivation in us,
who have inspired us
to dream
and fulfil our potential.
We thank you
for being our role models
to help us become who we are today.
I now bring this little poem to a close
with a chorus from our School Song:
Girls united from all nations,
Join we hands in common trust,
Strong of heart and pure in actions,
Thus we'll meet life's sorrow thrust.

October 2019

A Tribute to the Orange Party
In Celebration of Its 5th Birthday

And so they came,
my very dear friends,
brilliant men and women,
one couple after another,
a quartet,
to celebrate
the one and only Orange battle
bitterly fought
and narrowly won.
And so the first citric seed was sowed,
unplanned,
unintended,
almost accidentally...
yet the Tree of Friendship took root,
almost instantly,
with such camaraderie
even in the most mundane things we do,
as we look at the world together,
in words,
or picturesque albums,
or dance
or songs
or even poetry,
among much laughter, gaiety, hilarity
and good old joking and teasing,
while enjoying fine dining together,
and drinking tea or coffee,
 – water for one –
and fine wine for most others...
And like all orange trees,
we blossom in Spring,
thrive in Summer,
bearing fruit in Autumn,
enjoying the fruit of our labour of love in Winter...
and then blossoming in Spring again,

in full cycles...
And five years on,
the tree now stands tall,
blooming,
bursting with health and energy,
and bearing
the most wholesome Oranges
one will ever see,
a total of 8,
each as sweet and juicy and substantial
as the next,
similar and yet so different,
we need a code to know which is which:
A is 'State Leader'
with a vast pool of insider knowledge and secrets,
a walking encyclopaedia
with off-the-cuff
facts, figures, statistics
and intelligence
one cannot even find from Wikipedia;
C is 'Wen Zong',
a humourless serial entrepreneur and lawyer,
and a hopeless quiet behind-the-scenes worker;
E is 'Cheung Kong No. 1',
a superwoman
and a 24-hour multi-tasking champion
with a BB surgically attached
since time immemorial,
a paper tiger with a tender heart,
filled with love and compassion;
F is 'Thunder No. 1',
the most fun-loving doctor
who turns all his hobbies
into professional pursuits
and one of the kindest and generous men on earth;

H is 'North Point No. 1',
a well-read intellectual,
classy and sophisticated,
generous and fun loving,
the epitome
of knowing how to live life to the fullest;
J is 'Rose',
the poor lonely left-alone wife
with a workaholic husband
and has therefore
got into an annoying habit
of scribbling whatnots on paper;
R is 'Sunkist No. 1',
a model citizen
who spends almost every moment
on volunteer works,
and the artistic Supremo
with sharp eyes and magic fingers;
S is 'Zhong Shan No. 1',
the ultimate understatement,
who underplays his talents
and laughs at himself,
yet he's a baritone,
a keen pilot,
a seasoned scuba diver,
the supreme wine connoisseur
and everything else.

And so on this our special day,
let's celebrate our 5th birthday together,
let's recapture all those fun and laughter
we have shared together;
let's continue to nurture
this true friendship
that has stood the test of time,

let's pray that we may have more of such joy,
a joy that's so simple,
easy,
undemanding,
and yet so dear and warm,
loving and caring,
like a family,
till the end of time...

May 2017

A Thank You Tribute to My Comrades
A Beautiful Wedding Present for My Young Couple

And so he leaves me
and his father,
to join his bride,
in a love nest
5 minutes from ours,
and 3 if he runs
fast enough,
but still too far away
for my aching
but joyous heart
only a mother can feel and endure...
And so my friends came,
distinguished men and women,
13 of them,
to celebrate
a union
that could only come from Heaven,
and brought me,
nay,
the newly weds,
the Swarovski
Rainbow Fish Family,
in glorious splendour
of gold, yellow, purple, blue and shocking pink,
all swimming in one direction,
melting my heart
with such daunting spirit
and single-minded togetherness...
They also gave laisee,
and other whatnots;
One brought a model Ferrari,
and encouraged our son
to get a real one before long,
an ambitious challenge
for any youngster,

just out there in the rat race,
fighting with brutal bruises,
uninvited and oft unjust,
and rarely attended nor soothed.
And at last we sat down,
as we poured champagne
and waited for its bubbles to rise,
like thousands of sparkling smiling eyes –
and indeed we did smile a little too much
for we almost forgot dinner,
yet no one cared
for when "LOVE sits down to the banquet,
LOVE sits long"...
Then there were the quizzes,
all on wedding traditions and ancient beliefs.
No, how could one ever know
in time long gone by,
that 'bride' meant 'to cook'
and that Queen Victoria's wedding cake
weighed 300 lbs
or that she single-handedly
started the trend of wearing wedding dress in white.
Or that wedding rings
should be worn on the 4th finger of the left hand
because it's believed that
the vein in that hand
led directly to the heart;
and that no pearl
should be used
for wedding rings
as it has the shape of a tear.
But then all my learned friends
knew the answers
and much more,
and deserved

the prizes
of forks and spoons
with china handles in painted floral patterns
bought from Korea.

And so my darling boy,
I am so glad
you could make it
at long last,
and unwrap for yourself
the bright shining crystal light
from our dear friends.
And so from today,
let this light bless you and your bride,
and with these our friends
let it bless you two
that you may live a life
of health,
in joy and with purpose
forever and ever and ever.

September 2016

For All My Mediator Friends with Love

They gave me a 10 pounder –
with two birds
gazing at each other,
through a hollow-out centre
from an exquisite circle,
the colour of corals,
in deep orange-red,
in translucent crystal,
filled with free moving lustre of sheen,
of wild roses in adolescent bloom,
resembling a halo
of love, friendship and camaraderie –
and rightly entitled
"Sweet Love",
that we may be reminded
to cherish
all those around us,
when we never seem to have time for them,
yet all these joy and fulfilment,
are just for our taking,
if only we care to.

They gave me a 10 pounder,
these my mediator friends.
They created a ground rule not
found in the Code of Conduct,
and so what does a professional do
when her peers do not follow its Ps and Qs,
and when they are seasoned professionals
and teachers of the young and aspiring to carry our torch?
Then in the quiet of self-connection,
I hear the two birds in whisper sing,
strange outpouring of words of wisdom,
surrealist,
almost absurd:

enjoy, enjoy, enjoy
and so they keep singing...
And so in humility I learn –
enjoy this sweet moment I shall,
and to thank you for the gift
and what it means behind.
But still I need to say and hope you will remember
that the greatest gift is YOU.

July 2014

Farewell My Friends, Fare Thee Well...

For almost four years,
I laboured on this my poetry column,
pouring out my heart to you –
everything and anything
that have touched me,
from the ridiculous to the sublime;
from the tragic to larger than life;
from idle thoughts to deep contemplation;
from art to fashion;
from home to outer space;
from statesmen to supermen
to ordinary souls
who have made a difference
to our lives...

For more than three years,
I have given you my best,
laughing and crying with you –
I have celebrated lives
regardless of race or creed,
and wept for the deprived and the abused;
I have paid tribute
to those who have given us hope
and admonished the vicious and the wicked
who have fouled the air we breathe in.
I mocked the bizarre and the burlesque,
wrote on scandals caught on tape,
and lamented all those young lives
lost only too soon.

For more than three years,
I have been in there,
and I have been in there for too long –
I need to go and open the door.
Yes, I need to go and open the door,

just to see if there is a rose,
or a tree,
or a garden
on the other side...
I just need to go and see,
even if there is only a veil of darkness,
or the echo of the wind,
or nothing at all,
yet I need to venture out
and experience it for myself.
There are so many new challenges yet to conquer.
There are dreams to dream
and hopes to hope.
There are stars to embrace,
fires to kindle
and wonders to marvel at.
I want to fill my soul
with awe and beauty –
I want to look for that golden ladder
that I might climb up to the heavens...
And so,
this is the time to take leave
and say goodbye
for newer paths.
For all good things must come to an end,
just like when we enjoy the warmth of the sun,
we know it will soon fade
beneath the horizon.
Yes, good things have their comes and goes,
just as we see the most beautiful flower
that too must wither
and that it is then
when we will find it the sweetest,
that we know for sure
that the end of something good today

just means that something better
might come tomorrow,
and we shall be looking
for that rainbow
in the full glory of their colours.
So goodbye for now,
my friends,
I promise I will come back somehow.
Farewell,
my friends,
you whom I have never met
but you who must have known me a little by now.
Fare thee well,
my dear friends,
I will return to you somehow.
I hope some of my rhymes
will be in your memories
and when you perchance
remember a line or two
and think of me,
I will be there for you.
Farewell for now,
I will someday fulfil that vow,
I don't know when,
but I think we will meet again
sooner than we think.
And before we do,
may you always have walls for the winds,
a roof for the rain,
tea beside your TV set,
laughter to cheer you,
those you love near you
and what you desire most in your heart.
This is really the best parting
for such a humble beginning.

Some of my poems may not rhyme to your tune,
nor start or end in the way you wish,
let's just accept that nothing is perfect,
and life is about not knowing,
having to change,
taking the moment
and making the best of it,
without knowing what's going to happen next...
So adieu,
my friends,
parting is never easy
and I am so grateful
to have something
that makes saying goodbye so hard.
Goodbye once again and fare thee well.
Please remember,
goodbyes are not forever,
goodbyes are not the end;
they simply mean I'll miss you,
till we meet again...

December 2011

Ode to Knowles Building

We were but 18,
the Chosen Ones,
each with an entry ticket
to the elite world of tomorrow...
The getting in was difficult
but the staying in
almost painless...
Student loans and subsidies
were aplenty
when all's needed
was a proof of financial want...
Life was an autumn breeze,
a lazy stroll,
a warm afternoon –
there's always time
for sunshine
and moonlight,
a little romance,
and the not-so-clean Lily Pond,
long lunches at Czarina,
little harmless gossip after a tutorial,
staging a play or two at the Loke Yew Hall,
walking down High Street for late night 'sweet soups',
subtle flirting here and there,
and as for those who lived in dorms,
they played harder
to get sport trophies for their halls,
and organized high-table dinners,
barn dances
and the all-important Annual Ball,
examples of the very trivia
that preoccupied our young minds...
Then April came,
all such activities stopped –
our carefree faces turned anxious:

everybody was hurrying,
running from dorms to libraries,
from lectures to tutorials,
from bus stops to Knowles Building,
with books or notes in hand,
'hammering' away...
boys unshaved,
shirts unbuttoned,
dashing around in slippers,
and girls put up their hair,
wore glasses,
rushed past in flat shoes –
looking attractive was hardly top priority –
we fought to get a seat in the packed library
or find the most sought after 'reference book'
lost even to the library register;
borrowed notes
from the brainy,
the diligent
and the unselfish;
regretful for skipping
the 8:30 am statistics lectures
and begged the class 'statisticians'
for a last minute revision...
yes, that's life at the University of Hong Kong
in the early 70s...
except for the enlightened few
who carried the burdens of our Motherland
on their young shoulders:
defending Chinese sovereignty over Diaoyutai,
studying the causes and impact of the Cultural Revolution,
debating dialectical materialism,
the Class Struggle,
and the official use of the Chinese language in Hong Kong...

Then the storm came,
without warning,
rudely awakening
the tender soul
still sheltered within the Ivory Tower.
The stock market boomed
and many speculators
became millionaires overnight,
with the Heng Seng Index
hitting a record high of 1775 points
in the early March of 1973 –
then just as street sweepers
and all and sundry
were still having their 'rice in shark's fin soup',
the index fell to 150 by the end of the year,
burning every player,
the professional and the amateur...
The moneyed, the daring and the greedy among us
had the first taste of the real world
while virtually everyone of us,
even the least worldly,
or the most protected,
had experienced one way or another,
through family members, relatives or friends,
the almost total loss of family fortune,
huge or small,
with some of those we knew
ending up
in the Castle Peak Psychiatric Hospital...
But the sun shone brightly
on that November morning
the year after,
when we were banged
with the mortarboard on the head

by our Vice-Chancellor
at the City hall...
The 150 of us,
wide-eyed and impressionable,
had long dispersed months ago,
into the dark abyss of the unknown:
many in Government and banking,
some in academia,
a few turned professionals
some in corporate enterprises,
and a handful became entrepreneurs and stroke it rich...
Everyone was busy
working and playing hard;
searching for the ideal job;
seeking self-improvement,
exploring the self;
contributing the little we could
to the society that had nurtured us so well;
and quietly but persistently
looking for 'the other half',
and so exhausted by night,
we simply dropped dead on the sofa,
watching "Enjoy Yourself Tonight",
the first major popular variety show then,
more for tension release
than intellectual stimulation of any sort.
On Sundays we got up early
to wait behind a long impatient queue,
so the family
could enjoy a noisy dim sum lunch
in a packed restaurant,
a must for all filial sons and daughters...
and all these
while we started
in earnest

small families of our own...

With Thatcher's falling down
the steps of the Great Hall of the People in 1983,
the 1997 refrain was heard in diverse voices everywhere –
in every classroom,
every restaurant,
every train or bus ride,
yes, on every occasion,
happy or sad,
academic or political,
serious or lighthearted
by intellectuals,
the rich and the powerful,
and the man-in-the street...
After the signing of the Joint Declaration in 1984,
the Transition got off the ground
and triggered the first wave of emigration.
The 'astronaut syndrome'
with the husbands physically separated
from their wives and children
was perhaps the saddest phenomenon
of which the consequences
were to be despairingly felt much later
with divorces, loss of jobs and homes.
Some emigrated to
Canada, Australia and the USA,
some took British nationality,
but many more stayed
to embrace the return to the Motherland.
Some formed opinion groups and think tanks
and joined political parties
with one of us
being appointed to the Exco.
Those who stayed in the Government

are now Secretaries,
Directors,
Auditor General
and senior officials,
serving the public
with dedication and commitment.
Others invested or worked
across the border in the Pearl River Delta,
and later in Beijing and Shanghai
and other parts of the Mainland as well.
And yet everything
seemed to be overshadowed
by an uncertainty of the future.
Many joined rallies
in support of the students
in Tiananmen Square.
The drafting of the Basic Law
took many years and much consultations
and the Handover Ceremony was dignified
as the whole world watched.
As history has shown,
"One Country Two Systems"
has worked well,
and Beijing has been supportive
of the HKSAR,
and many have since returned,
only to find house prices
have much risen and job opportunities gone,
but return they did.
This journey from colony to SAR
and from Borrowed Place Borrowed Time
to part of China
was historic,
eventful,
sometimes step-toed

but most importantly right
and not regretted.

The Reform and Opening Up
initiated by Deng Xiaoping
which began in 1979
changed the country and Hong Kong
in ways that none of us,
perhaps not even the wisest among us,
nor Comrade Deng himself,
would have foreseen.
Our ancient country,
burdened by dogma,
feudalistic traditions
and political strife,
and utterly destroyed
by the Great Proletarian Cultural Revolution,
blossomed with energy,
enterprise and pragmatism,
turning itself
from an agrarian economy
to the Workshop of the World
and now beyond,
leading the global move
to new energy cars
and other elements of sustainable development.
Foreign exchange reserve grew
from a pitiful, meagre US$2.2 billion in 1979
to US$2 trillion,
now largest in the world.
Hong Kong manufacturers
and other businessmen,
including some of us or their bosses
were the first to invest in China,
growing to become

leading world suppliers
of garments, electronics, toys
and other consumer products;
then diversifying into real estate, hotels and services.
Others became bankers,
bringing Chinese banks
onto the world stage,
flourishing
while Citi and the Bank of America
floundered.
As citizens of emerging China
we now stand tall.
Our state leaders
now sit at the top table at global summits,
our sovereign fund's investments are sought
in the cash-hungry post-financial tsunami world,
and ICBC
is now the largest bank
by market capitalization in the world.
The Beijing Olympics
was painstakingly organized
with the Zhang Yimou opening gala
a *tour de force* of the Modern China.
The Sichuan Earthquake
was a natural disaster
that brought the Chinese people,
from prime minister to peasant,
and at home and abroad,
together.
We redefined our self-identity
from Hong Kong People from Hong Kong
to Chinese People from Hong Kong,
and more recently,
to Hong Kong People of China –
a rite of passage

that was transformational,
enlightened
and profound.

We grew up
listening to Elvis and Beatles,
Peter, Paul and Mary,
Tom Jones and Joan Baez,
and knew little local lyrics.
The then home-grown stars
like the Wynners,
Sam Hui,
Irene Ryder,
and Teddy Robin and the Playboys
sang mostly English songs too.
After we left the University,
Cantonese songs
came into vogue
thanks to the popularizing
of the walkman,
TV soap opera series
and talented lyricists and song writers
like Wong Jim,
Cheng Kwok Kwong
and Koo Ka Fai
who were creating immortal pieces
that were sung and heard everyday
on radio and TV programmes
turning Roman Tam, Anita Mui and Leslie Cheung,
all now no longer with us,
the biggest of Hong Kong popstars.
Domestic movies in the late 80s
would put Hong Kong cinema
on the international map.
Jackie Chan excites us

with his real 'kung fu'
while Chow Yun-fat,
in collaboration
with Director John Woo,
sets the *de facto* standards for triad films
in "A Better Tomorrow".
The "Miss Hong Kong Pageant"
debuted with much fanfare and excitement
but has over the years
become an annual non-event.
For leisure reading
girls moved
from our adolescent Taiwanese heroine,
King Yiu,
to Yik Sue,
then Lee Big Kwun
and Leung Fung Yee,
while the boys remained loyal
to their Ngai Hong,
Kam Yung, Leung Yu Seng and Koo Lung.
Local journalism thrived too,
with the arrival of City Magazine,
pioneered by one of us,
a serious writer and philosopher.
The much acclaimed Economic Journal
joined mainstream media,
while sensational papers
like Apple and The Sun
also entered the race
and emerged as the most popular tabloids
for their scandalous reporting style.
Others such as the Eastern Express
came and went.
The once widely read Sing Pao
and Hong Kong Standard

are languishing
just as our beloved Daipai Dongs
are now
but historic relic...

The world outside
has seen its ups and downs.
Watergate was a bungled burglary
that took on a grand scale
and ended a presidency
and our innocence.
The Vietnam War drew to a close
with its ravage
immortalized by a photo
of a terrified, crying naked burnt girl
running for life with fire on her back
and that of the last helicopter
departing the roof
of the US Embassy
in the then Saigon
with fearful locals hanging
from its landing rails.
American parents
heaved a deep sigh of relief
that their sons
would no longer be needed to be sent
to fight for a war
that they did not understand.
Decades later,
the same fate was repeated
for young soldiers
dispatched to Afghanistan
and then Iraq,
risking or losing lives
for no just cause.

The World Trade Center in New York
fell on that fatal morning
on September 11, in 2001
and the United States of America
started the War Against Terror.
Governments came and went,
Tory and Labour,
Republicans and Democrats,
Reagan and Thatcher,
and Bill Clinton
with Bushes before and after
and Hillary and Monica in between.
Gorbachev got into and out of power,
and the Berlin Wall fell...
A Polish shipyard worker
became President
while another of his countryman
a Pope.
Mitterand died
with wife and mistress
both at his funeral
and Sarkozy changed First Ladies
while dealing
with the same Merkel.
But the whole world
is captivated
by the Obama magic,
with Michelle
being the most charismatic First Lady
of them all:
Harvard and Princeton educated,
lawyer,
mother of two
and a direct descendant
of black slaves,

she and her husband
exemplify the American Dream
and render their country
a great nation
again.

And so the world moves on,
from the Avian Flu in 1998
to SARS in 2003
to today's Swine Flu
with government
learning from past experiences
and fighting hard
to stop the spread
of this contagious,
mutating virus
that is inflicting millions globally,
and doing better
with our very own Dr Margaret Chan
as the head of WHO.
Taiwan has just taken part
in the World Health Assembly
for the first time
since the UN seat went to Beijing,
signifying a historic thaw
in cross-strait relationship.
Elsewhere gestures and postures prevailed.
North Korea
rejected appeals to rejoin talks
on its nuclear programme,
and Roh Moo-hyun,
the first South Korean President
to cross the 38th parallel,
committed suicide.
The Pope visited Jordan,

Israel and the Palestinian territories,
but failed to please his Israeli hosts
who wanted more forceful language
on the Holocaust.
The Italians
introduced a policy
of returning boatloads of migrants
to Libya
before any can claim asylum.
The financial tsunami struck in 2008,
Wall Street went broke,
and General Motors' share price
has fallen to its lowest level
since the Depression.
Yet rich Indonesians in Singapore
still line up outside Louis Vuitton
to get in,
the Gilt Groupe
provides access,
by invitation only,
to coveted fashion and luxury brands
with each sale lasting 36 hours
and featuring hand selected styles
from individual designers;
Japanese volunteers
from school to middle age,
in emerald-green vests
and mustard-colour cleaning gloves,
methodically cleaned up
the streets of Paris;
celebrity artist Damien Hirst
sold his "The Golden Calf"
for £10.3 m
just a day before Lehman Brothers
went bankrupt,

and Obama was awarded the Nobel Peace Prize
for his vision
after just nine months in power,
stunning the world
and leaving the US President
'surprised and humbled'.

And as for us,
the 74ers,
we soldier on
from shore to shore,
from job to job,
from crisis to crisis,
from year to year,
much enlightened,
wiser,
lighter,
and hopefully happier,
younger
healthier
and more fulfilled
by the day.

September 2009

A Tribute to My Gang of Four
To Sanny – Parting Is Such Sweet Sorrow

When Ross left us for his radio call,
And dwindled our "Gang of Four",
We felt so upset and distressed,
With a sadness that could only be suppressed.

Our little "English Game" we now put aside,
The list pinned on the board we no longer eyed,
How we could find it so funny we wondered why,
Life would never be the same, we wanted to cry.

Then came the ring asking for you, Liza and me,
Early in the morning just as we finished tea,
How he brightened our day with his cheerful tone,
As we listened in turn through the heated phone.

How he would just appear with his signature smile,
One hand the heavy recorder and the other his file,
As we stayed in our little corner we called our den,
Talked about old times over and over again.

So on this day when we have to say goodbye,
We shall not stop you, we shan't even try;
We just know that you would soon drop by,
In person, and if not, then through the line.

Let's together sweeten the sorrow parting brings,
And fill it with anticipation and all little things;
Share with us your new adventures as we dine,
In verses, serious or lighthearted, I'll send you mine.

But should you feel you are fighting all alone,
Grappling with endless problems on your own;
When there's no one to side with you,
Or back you up in whatever you wish to do;
When you grow weary and feel like to explode,

As you stumble along the dreary road –
Please just know that we are always here,
Ready to share with you our heart and ear.

October 1979

A Halo of Gentle Souls

A Tribute to Our Dearest Mary
Stillness of the Dawn

In the stillness of the dawn,
she woke up
to the gentle voice
from afar,
calling her in soft whispers,
caressing her ever so tenderly,
and serenading her
with songs of love and peace and beauty...
It came nearer and nearer,
in whispering words,
a father to his child,
beckoning her with His Hand of Mercy,
and she surrendered
into His loving arms,
with tears of joy in her waiting eyes –
for she has searched for Him,
through each corner of her soul,
for years and years,
and now her prayers were answered,
and He came for her
that she might reunite with her love
forever and forever.
She is a brave soldier of the Lord.

In the stillness of the dawn,
I woke up
to her gentle face,
a face
that has oft calmed the frightened,
with a smile
that has soothed many the sick,
and eyes
that have lit up the darkest of souls.
How she has listened
to the heart of the despondent;

provided for those
who are deprived, denied and in need,
attended to the ailing and the searching
who came to her for light...
but most of all,
she sang.
She sang
for the sick,
the elderly,
the lost,
the weak,
the needy,
the disheartened,
her songs of love and peace, hope and joy;
and she sang for her friends too,
during every bible class,
every visit to the poor and elderly,
every celebration,
every memorial service,
every gathering,
every chance,
every moment...
and she sang for me,
once on my special night,
she sang for me,
in her full cheongsam
she sang for me,
when she should be resting her voice...
and I cried then,
I cried the cries of a child
so much loved and spoiled by a doting mother...
and now as she is in heaven
with my heart shredded into a thousand pieces,
I cried.
I cried the cries of a mentee

who misses her mentor,
I cried the cries of a stubborn soul
who said no to her repeated urge
that I should return,
I cried the cries of a friend
who has not gotten to say good-bye...
yet I know,
deep in my heart
I know,
that you will wait in heaven
from this moment on,
till God the Almighty
asks you
to bring me home.

January 2014

To Suk Suk,
Our Dearest Uncle and Mentor

You never said you were leaving,
you never said good-bye;
you were gone before we knew it,
leaving us in total despair,
with a grief so deep,
a loss so sudden
and a mourning
too heart breaking to bear...
And if grief could build a stairway
and sorrow a lane,
I'd walk right up to Heaven
and bring you home again.

No parting words were uttered,
"I like to take a rest" you said after breakfast,
then you left with the bottle of Yakult still tight in your hand.
We had but lunched together only days ago.
How you liked the new home,
surrounded by the clear blue sea –
it's selected specially for you,
yet you stayed for only five days.
I know you wouldn't want to see us in pieces,
but the loss is too new,
and the grief too grave,
so please just let us be...

I have known you even before I joined the family –
How oft have I been told
of the days
when Carson was in Singapore
when he was only 17;
and how you and Ah Yee
took such great care of him,
like he's your very own.
You bought him his first watch

which he has kept until today.
There's the forever Sunday buffet lunch,
the outdoor picnic,
the whirlwind drive,
the cycling,
and all the intellectual debates
on social, economic and political issues;
and the delightful discussions and sharing
on all things under the sun.
"Suk Suk is like a father to me",
he used to tell me,
"my role model
of what a father and husband,
and indeed a man should be."
And how you welcomed me
with open arms
into the family,
and I recognized soon enough,
back then,
the strength you possessed,
as measured
by the generosity of your heart:
it's all about your quiet giving and loving,
that spellbinding radiation
of pure and unconditional love,
not only to your immediate family,
but to all of us
who are fortunate
to be a part of it.
You and Ah Yee
were the very glue
that tied us together,
starting a Sunday family lunch tradition
for almost thirty years,
touching the lives of many,

and that of mine,
teaching even the young ones
to realize that
with sms, twitter, or skype,
face-to-face communication still trumps them all.
You were always so gentle,
so loving and kind,
patiently listening
to all our silly nonsense
and giving advice only when solicited,
and what sound and sensible advice you always gave!
Never have I heard a loud or strong word
from you,
nor seen your brows knitted with negative emotions,
for nothing could fluster your genteel soul,
not when you alone lost to the other threes
on a mah jong table,
nor when you were behind the wheel,
taking us to every new restaurant discovered,
driving like a young man,
even when you were over 70,
fast but safe.
And so when all the tears are shed,
and when we think of you,
we remember a life well lived,
a life of diligence,
dignity,
devotion,
love and compassion,
and all the reasons we love you;
and know with our heart
why God himself would want you by His side.
If roses grow in Heaven,
Lord please pick a bunch for me,
and place them in Suk Suk's arms

and tell him that they are from me.
Tell him I love and miss him
and he holds a special place in my heart
that nobody could ever fill.
It broke my heart
the day he left,
but tell him he did not go alone
for a part of me went with him,
the day when You took him home.

May 2012

In Loving Memory of Our Dearest Friend
An Exemplary Son of Hong Kong

And so
he left us,
quietly,
without warning,
returning home to his Heavenly Father,
leaving us in total despair,
with a grief so deep,
a loss so sudden,
and a mourning
too heartbreaking to bear....
But we have to let go,
for he must be very tired,
having lived a life almost without himself,
devoted in selfless service to the community
in numerous capacities for more than twenty years....
Now he is with The Lord,
where he is resting
at last,
in peace.

How many times
he has spoken
the words
we feel deep in our hearts
but hesitant to say;
how he has,
with courage,
totally unafraid,
spoken the truth,
when others,
even the honest,
the just,
and the fair,
have but kept silent;
how he has triumphed

even as he's jeered at
by his 'dog biscuits'
and 'loss of memory'.
Now tributes flow in
not only from friends and allies
but his opponents,
for his relentless efforts
in helping to draft the Basic Law,
in his support of "One Country, Two Systems"
and his contribution to the smooth transition of Hong Kong.
A role model
for all and especially the young,
he is a medical doctor with compassion,
a thinker with insight and an independent mind,
a social critic with a conscience,
a politician with a soul,
a man with character,
and,
a friend with a heart.
Yes, we are blessed
to be one of his friends,
for to them,
he gives selflessly.
We know,
we have experienced his kindness
firsthand,
again and again.
We shall always treasure
the wisdom he has shared with us,
and in particular,
his insight of what we should do
to make Hong Kong a better place,
and his love for the Motherland...

Dearest Raymond,

our friend and mentor,
we shall miss you
forever!

November 2007

Celebrating a Life Well Lived
A Man who has Lived a Full Life

"Daddy wen to heaven at about 3 am this morning"
the 't' is missing
"there could never be
any spelling mistakes,"
she once told me,
"the computer
checks the mistakes
for you in red." –
but this is a time of bereavement,
the loss is too new,
the grief too grave –
yet her faith is strong,
and her conviction firm:
daddy has lived a full life,
and now that he has done
what he has done,
his body
is leaving the earth
and his soul returning home
to The Lord,
smiling,
rejoicing...

White flowers
of every possible kind
doubly stacked
along the three walls,
the entrance,
the hallways...
I could not find the one I sent –
I want the guests to see
the verses
I chose for him:
but it will not be necessary
wherever it may be –

the verses
are visible everywhere
as we enter the hall –
"May your deeds be shown to your servants,
your splendour to their children." Psalm 90:16.
We see them
in his children,
his grandchildren,
his relatives and friends,
the young and the elderly
from all over the world
gathering together
to celebrate
a life well lived...

Yes, we are indeed celebrating –
we are celebrating
the life of a 92-year-old,
a life of diligence,
dignity,
devotion,
love and compassion
and
a life full of God –
as his young grandchild
stands up to say:
"Yeh Yeh always puts God before everything else."

Yes, we see his life
in his five daughters and two sons,
his 13 grandchildren,
and thousands or more,
whom he has touched,
some way,
some place,

some time.
And when the 'family choir'
begin to sing
"All Things Bright and Beautiful",
I experience
'how great is God Almighty'
and
my heart weeps
with joy
for a dear friend
who sings with such love
and inner peace,
reconciled
with God's plan
and
rejoiced
that their loved one
is finally home
with his Heavenly Father...
free at last
from all earthly
pain and suffering.

"She has been a friend"
that's all I want
on my tombstone,"
she once confided in me
years ago...
When one is blessed
with an earthly father
like hers,
and the faith of the Heavenly Father,
no stone is large enough
to write down
all that

she has offered
to all those
she has touched
some way,
some place
some time...

Hers will also be
another life well lived.

April 2006

Light a Candle

Light a candle for him,
our young friend,
pride of his family
and joy to his friends –
how he had touched the hearts
of so many
who had crossed his path
with his gentle kindness and compassion;
how he had inspired us
with his sensitive writing;
and the music
he had created
to awaken this fallen world...
but then the tiger came to him at night,
enraging him with its seething claws,
provoking him with its thundering roars,
and exasperating him with its arrogant stance,
turning him angry,
infuriatingly mad,
and beside himself...
For years he had suffered
with an anguish that could not be explained,
a pain that could not be spoken,
and a grief that could not be consoled,
but he just fought on,
this our young soldier,
he simply fought on,
for he was proud,
he was indomitable,
he was unyielding,
all on his own,
allowing no help,
soliciting no advice,
one year after another,
our fighter battled on,

gradually withering away,
until he's no more...
Our Heavenly Father
nurtured him with tender love and care
and through Him
he found peace and tranquillity
as He carried him home
in his loving arms...

Light a candle for him,
this our young friend,
for into a new life,
he is now reborn.
He has left his mortal prison
to enter a new world
where there is neither darkness,
nor pain,
nor suffering,
nor grief,
nor anguish,
nor fear,
nor despair,
nor injustice...
Yes, he needs fight no more –
the shadow of darkness
has now forever slipped away from him,
for the Light of his Heavenly Father
has spread over the dark nooks of his soul,
that he might see light.
Let's light a candle
for our young prince,
and be relieved to know
that he has,
at long last,
found peace,

and is at rest now,
enjoying
the splendour of eternity and peace.
Yes, let's light a candle
for our young warrior,
let's weave his garland
with flowers of everlasting beauty,
and crown his soul with coloured light
of freedom and peace.
Let's not mourn for him no more,
for he is in a much better place
where we would all wish to go to,
one day,
where we will join him
and all our loved ones
who have gone before us,
in harmony and peace...

January 2006

Written by: Julia Wen
Edited by: Betty Wong
Cover designed by: Jimmy Fong
Illustrations by: Ruth Lau
Published by:
The Commercial Press (H.K.) Ltd.
8/F, Eastern Central Plaza, 3 Yiu Hing Road,
Shau Kei Wan, Hong Kong
Distributed by:
The SUP Publishing Logistics (H.K.) Ltd.,
16/F, Tsuen Wan Industrial Centre,
220 –248 Texaco Road, Tsuen Wan,
NT, Hong Kong
Printed by:
Elegance Printing and Book Binding Co. Ltd.
Block A, 4th Floor, Hoi Bun Building
6 Wing Yip Street, Kwun Tong, Kowloon, Hong Kong
© 2021 The Commercial Press (H.K.) Ltd.
First Edition, First Printing, October 2021
ISBN 978 962 07 0593 9

Printed in Hong Kong